Education Division

Held at the 2024 AIChE Annual Meeting

San Diego, California, USA
27-31 October 2024

ISBN: 979-8-3313-1661-7

Printed from e-media with permission by:

Curran Associates, Inc.
57 Morehouse Lane
Red Hook, NY 12571

Some format issues inherent in the e-media version may also appear in this print version.

AIChE
120 Wall Street, FL 23
New York, NY 10005-4020

Phone: (800) 242-4363
Fax: (203) 775-5177

www.aiche.org

Additional copies of this publication are available from:

Curran Associates, Inc.
57 Morehouse Lane
Red Hook, NY 12571 USA
Phone: 845-758-0400
Fax: 845-758-2633
Email: curran@proceedings.com
Web: www.proceedings.com

TABLE OF CONTENTS

CORE CHE CURRICULUM COURSES

ELECTIVE AND CROSS-DISCIPLINARY COURSES IN CHE

COMPUTING, DATA SCIENCE, AND NUMERICAL METHODS IN THE CHE CURRICULUM

ASSESSMENT METHODS

STUDENT SUPPORT AND ADVISING

FACULTY PROFESSIONAL DEVELOPMENT - WORKSHOP AND TALKS

TEACHING TIPS

ENGINEERING OUTREACH

DEPARTMENT HEADS FORUM (INVITED TALKS)

EDUCATION DIVISION AWARDS PLENARY (INVITED TALKS)

IN HONOR OF PROFESSOR JEFFREY A. REIMER, THE 2023 RECIPIENT OF THE WARREN K. LEWIS AWARD (INVITED TALKS)

INTRODUCTION TO NSF'S CBET DIVISION AND PROPOSAL WRITING STRATEGIES

NEW FACULTY FORUM (INVITED TALKS)

NEW FUNDING OPPORTUNITIES AT NSF AND PROGRAM OFFICER PANEL DISCUSSION

POSTER SESSION: CHEMICAL ENGINEERING EDUCATION

TEACHING SURVEY RESULTS (INVITED TALKS)

ENGINEERING EDUCATION RESEARCH

SUSTAINABILITY

ENGAGEMENT AND INNOVATION IN EMERGING LABORATORY TOPICS

TEACHING PROFESSIONAL SKILLS

57a Integrating Creativity and Innovation While Utilizing Local Materials in Undergraduate Engineering Design Curriculum

Katrina J. Donovan, Jon J. Kellar, Michael K. West

Creation of pottery and ceramics is as ancient as human settlement. The local minerals found in the Black Hills have been used to make ceramics for hundreds of years. The talk focuses on leveraging the local resources to develop hands on curriculum to inspire students in a curricular setting to be more innovative with their design components. The talk will present case studies of freshman first year experience courses and a junior level design course. A rigorous external assessment of the research will be conducted and includes the use of a variety of assessment tools a couple of the following will be used: Herrmann Brain Dominance Inventory, Small Group Instructional Diagnosis (SGID), student focus groups.

Results from the SGID and student focus group surveys report positive results with the modified curricula that has been integrated into the initial first year engineering course. Student surveys were performed with a primary focus on student advancement in areas of creativity, innovation, and technical knowledge. The self-efficacy studies illustrate a general increase in the students' perception of their creative skills and technical knowledge.

57c Designing Engineering Case Studies to Practice Open-Ended Problem Solving in Core Courses

Camille Bilodeau

Most chemical engineering core courses focus on solving problems that have a clear right and wrong answer. Despite this, after graduation most students will need to learn to utilize the skills they have learned in class to solve poorly defined or open-ended problems. Here, we explore the use of in-class case studies to enable students to practice applying course material to open-ended problems. Specifically, our case studies are inspired by case studies in business school settings and are designed to practice decision making and teamwork, where the decisions are informed by the technical concepts taught in the course. We developed three application-oriented case studies for a sophomore-level numerical methods course (which is a core course in our department). We will discuss the core components and design parameters of these case studies and how they qualitatively impact learning objectives (evaluated using interview and survey data). Finally, we will discuss how this methodology can be used in the context of chemical engineering core/elective courses more generally.

57d Engaging Students through History: How Thermodynamics' Past Motivations Transcends through Time

Jennifer Kao, Glaucia Carvalho Do Prado

Thermodynamics is a fundamental core engineering course and considered by students one of the most challenging courses in the engineering curriculum. Moreover, studies suggest that engineering classes do little to foster conceptual understanding and very often students memorize the steps to solve a problem instead of understanding the reason and the concepts involved in the solution. Anecdotally, many students come to thermodynamics classes with fear and a preconceived mindset that they will not be able to follow what will be taught. Additionally, many university professors agree that it is challenging to engage students in disciplines that are excessively mathematical and disconnected from human interaction. While many teaching practices have been reported to engage students in thermodynamics courses and connect it with real world examples, there are only a few instances that explore the connection of thermodynamics with its origins. Through the lens of the founding fathers of thermodynamics concepts, it is hypothesized that by incorporating historical elements in the thermodynamics course (such as the motivation and the thinking process for the scientists to come up with the equations students are learning), students will be able to make connections between the material and its relationship and importance within society, politics, and economics. This may serve as motivation to change students' mindset towards thermodynamics as well as promote enduring understanding that will facilitate retention of the concepts presented in class.

The history of thermodynamics was included for the first time in the chemical engineering thermodynamics course AT THE INSTITUTION in 2023. The instructor prepared a block flow chart containing the motivation, results and conclusions for each equation, process, and/or concepts mentioned in the course. One example includes Carl Lindes' motivation to develop the liquefaction process on a large scale after being requested by Guinness to supply liquid carbon dioxide to improve their beers' foam head. A voluntary survey was conducted at the end of the course to evaluate students' perceptions of the thermodynamics course. Results indicate that students found the thermodynamics course interesting (4.1 +/- 0.8 on a 5.0 scale) and that the history of thermodynamics presented in class increased their interest in thermodynamics (3.5 +/- 1.1 on a 5.0 scale).

With the positive feedback received by students, a new interactive website about the history of thermodynamics was developed and the effect of the website on students' retention of thermodynamics concepts is being currently investigated. Chemical engineering students who attended thermodynamics course AT THE INSTITUTION in 2024 were assigned as the control group and were not provided with any historical background in class as well as an interactive timeline as a resource. A mandatory activity was conducted at the last day of instruction where students were asked to answer conceptual questions about thermodynamics. Students received credit (2% of final grade) for completion regardless of if their answers were correct. The new website containing the historical background as well as the interactive timeline will be included in the next offering of thermodynamics course in the chemical engineering department AT THE INSTITUTION (Winter 2025) by the same instructor. Students will be asked to answer the same conceptual questions and the results will be compared to the control group, along with their overall average grade.

Therefore, the objective of this work in progress is to present the interactive website containing the history of thermodynamics for those interested in including it in their thermodynamics course and present the partial results. It is expected that the outcome of this study will provide information on the possibility of using a novel and creative teaching practice that may (i) help students understand the concepts studied in a complex engineering course, (ii) improve the attractiveness and motivation of students to study a complex subject such as thermodynamics through focusing on the historical elements of "when, who, what, and where" and combining with the engineering aspects to solve the "why, how, and made-from what", and (iii) promote social awareness and the connection between the topics studied in thermodynamics with past and current economics, politics, and intellectual environment, which are crucial factors that determine how current world problems can be solved by engineers.

57e Development and Uptake of a Second-Year Chemical Engineering Design Open Textbook

Jonathan Verrett

Open educational resources (OER) have been shown to benefit students and faculty in several ways. For students, OER can remove barriers which may be financial, time-limited access, or resource format. For faculty, OER can be adapted to their course needs, for example taking or revising portions of the work based on the licensing.

This work outlines the development and impact of a textbook for a sophomore introduction to chemical engineering design course. The textbook parallels the design course and showcases chemical engineering concepts at an introductory level. The textbook consists of sections including chemical processes and diagrams, reaction chemistry and kinetics, phase equilibrium (separations), energy balances, unsteady-state operations & process control. The textbook has been active and in use since the fall of 2020. User statistics have been collected and results from these statistics will be presented in terms of country of access, device accessing the site, number of interactions, downloads and the development of these statistics over time.

57f Advantages-Disadvantages of Interdisciplinary Capstone Design for Chemical Engineers

Greg Ogden, Kimberly Ogden

This paper will discuss the evolving aspects of having chemical engineering students participate in college of engineering-based capstone senior design projects. Many schools currently offer hands-on widget-based capstone design projects emphasizing teamwork, collaboration and communication-skills needed by industry. While ChE programs offer traditional paper-based design projects focus on process design at the plant/facility level where plant costs are in the millions, if not hundreds of millions of dollars. This scale is in drastic contrast of widget-based projects with capital costs under $10,000. Chemical Engineering Departments looking to participate in college-wide Capstone courses need to evaluate opportunities and how the capstone courses meet ChE ABET criteria. Although it might appear to be easier to meet the social, economic, global and economic criteria related to ABET in a joint capstone course, many programs typically work to attain these requirements in program-specific courses. There is a subtle balance that can be achieved between participating in college-wide capstone courses, the desire for ChE students to participate in a hands-on build project and meeting ABET requirements.

121a "Synthetic Fuel Production" As a Chemical Engineering Energy Elective

John Kuhn

While many chemical engineers work in fossil fuel industries, we are also at the forefront of technology in renewable energy. With students capable of taking many pathways in other majors, it is important that we recruit and educate students in chemical engineering about the breadth of possible careers and its global impact. In our department, we have stressed environmental implications in several courses, with topics on solar engineering, sustainability, and energy from waste that have been offered as upper level and graduate electives. A recent new course is "Synthetic Fuel Production", which focuses on the chemical and biological processing of both fossil and renewable feedstocks. The course also discusses current and futuristic fuel economies, such as hydrogen and methanol. In addition, the course is designed to emphasize the global aspects of both the history of energy, specifically transportation fuels, and related geopolitical impacts, current events, and potential advances for the future. With the broad scope and application framework, the course is capable to engage students within applications, while introducing/reinforcing core course content. It is part of USF's Global Citizen Project.

121b Drug Development and Pharmacology As an Upper-Level Elective for Engineering Pre-Med Students

Tanyel Kiziltepe

One of the challenges in a Chemical Engineering program is the development of a pre-med curriculum that must integrate well into the traditional chemical engineering curricula while still preparing the pre-med students with the necessary foundation for successful outcomes in medical schools. Most students who choose the pre-med track in a chemical engineering department are 1) either uncertain about pursuing a career between chemical engineering or medical field, and would like to learn more before committing to one, 2) they enjoy engineering approaches over basic sciences such as biology, and believe that the problem solving skills and technology savviness developed in an engineering department will differentiate them among their peers in the medical school environment, or 3) would like to pursue an MD/PhD graduate program where they can simultaneously contribute to society on the bench side as PhD engineers as well as on the bed side as clinicians. Developing such a training program must take considerations that go well beyond the completion of bachelor's studies for a chemical engineering degree while simultaneously meeting all criteria required for preparation and acceptance into medical schools. This challenge becomes most apparent during the identification and development of upper-level electives that can fulfill the tech elective criteria of the engineering department while providing students with the necessary foundation for medical schools. Here, the strategy to develop a course on Drug Development and Pharmacology will be described. The course stretches over several multidisciplinary topics including: 1) the drug development from discovery methods through clinical trials, 2) basic principles of pharmacology including ADME, pharmacodynamics, pharmacokinetics, and drug metabolism, 3) dives deep into literature to discuss cutting edge research pursued by chemical and biological engineers in the field of rational drug design as well as biomolecular engineering solutions to circumvent pharmacology related problems.

DDP is offered as a one semester upper-level (senior/graduate-level) course at Notre Dame, so that both graduate students, who find it complementary to their PhD studies, and the pre-med undergraduate students can benefit from it. As for the prerequisites, typically, one organic chemistry, one basic biology course with an emphasis on cellular systems is adequate. The course is divided into 3 major modules where each covers a critical fundamental component of DDP followed by a relevant literature example of how engineers approach drug discovery and development and in designing solutions to issues relating to human health. The assessment of the students involves homework assignments, exams, literature presentations, and a final term project where the students work on teams on a topic of their interest.

This presentation outlines the approach used to develop a DDP course in a Chemical and Biomolecular Engineering Department that meets the tech elective criteria while at the same time provide the pre-med engineering students with a foundation in drug development process and pharmacology to help them succeed in medical school. This class has been proven to be very beneficial for not only the pre-med students, but also for those interested in pursuing a career in pharmaceutical industry, medical device companies, medical or pharma arm of consulting firms, as well as those planning of pursuing a PhD in a Biological Engineering related field. In the past 8 years this class has been offered, it has received very favorable feedback particularly from the recent graduates in graduate or medical schools who wanted to make a point in reaching out to us to let us know about their appreciation as the teachings significantly benefited them during their ongoing studies.

121c Community Based Learning in Sustainability Engineering Education

Andrea Hicks

Introduction: Industrial ecology and sustainability education have evolved over time, as this relatively new discipline has emerged and has been refined. Courses engaging with these topics are often found in engineering departments, and feature project or problem-based learning (PBL) commonly with an industrial partner. This provides students the opportunity to apply the topics that they have learned in class in a real-world situation, complete with the constraints and uncertainties inherent in practice. Community-based learning (CBL), a combination of service learning and PBL, allows students to not only apply the material from a sustainability course, but also to do so in service of a community partner. This work presents an analysis of a multi-year effort in teaching a CBL course on the topic of industrial ecology and sustainability in a North American context.

Although often IE and LCA courses and curriculum are centered around industry partnerships, there is also the potential to focus them using a CBL or service-learning structure. Problem and PBL have been utilized to teach IE and LCA, typically with industrial partners, in the context of "learning by doing", where students complete a project with uncertainties and unknowns for an industrial partner. Potential exists to combine these two types of "learning through doing," to provide students with the opportunity to learn about IE and LCA through engaging with a community partner. Utilizing Brundiers et al. (2010)'s definitions of PBL and service-learning, the author would posit that CBL is a combination of the two, retaining the characteristics of PBL in the collaboration with partners, co-production of knowledge, supervision by a faculty and stakeholder, and systemic innovation with service to the community, with also the characteristics of service learning such as educating people and working in support of societal change. CBL courses often connect students with a partnering community organization, where the students' learning experience is enriched through working with the partner, and the students' work serves the partner in some fashion. CBL has been defined by the Morgridge Center for Public Service at the University of Wisconsin–Madison as "a credit-bearing educational experience that integrates meaningful community service with guided reflection to enhance students' understanding of course content as well as their sense of civic responsibility". Although these courses are commonly considered outside the purview of an engineering (and IE) education, there is a significant and often untapped potential for students to enrich their knowledge of IE through working with and in service to a community partner, and the potential for co-creation of knowledge with that partner.

The notion of combining PBL and service-learning courses is not new, although the CBL terminology may be with respect to engineering education. The field of chemistry provides ample examples of this occurring in STEM (Science, Technology, Engineering, and Math) education. Draper (2004) describes the coupled "project-based service-learning" integration in an advanced environmental chemistry course, and a statistically significant increase in the average score of student responses to end of semester evaluations compared to previous conventional offerings, suggesting that students may enjoy the ability to serve the community while learning, and the potential for it to give greater meaning to their efforts.

Methods: Each semester students work in teams (ideally in groups of three to four students, although this has changed over the iterative offerings of the course) to evaluate the environmental, economic, and societal considerations of a product or process identified by their community partner as a critical issue for which assistance is desired. The teams utilize tools such as LCA, material flow analysis, economic valuation, and societal considerations, that had been introduced and explored during the course. They are tasked with providing a recommendation to the partner based on the analyses performed. These tools and concepts are explored during the course in a typical lecture format with weekly assessments.

The project partners are typically procured as part of the UniverCITY year program at the University of Wisconsin–Madison, where each year the university partners with a new public partner (for a 3-year duration), who submit project proposals that they would like to collaborate on. Some of the particular projects completed over the years have been described previously in prior publications. The partners over the years have largely been different, including both urban and rural counties in the State of Wisconsin, such as Dane, Brown, Pepin, and Green County. As projects are identified by and defined with the partner, they have varied each year, although they have included phosphorus control at small-scale wastewater treatment plants, sustainable low-income housing materials, solar panels and wind turbines at a small school district, manure management from dairy cattle, and recovering blue-green algae from waterways to be used for a beneficial product. In general, the community partners have an interest in sustainability, but often lack the resources to conduct a quantitative assessment of the sustainability of a planned project or product acquisition.

Results: Qualitative data has been collected as the student experience in the course, and the "role of CBL in sustainability education." The reflections are aggregated into a word cloud. Two brief responses warrant paraphrased mention based on this reflection, first that it goes beyond idealized coursework problems which illustrates the PBL-based aspect of the project, and second, sustainability is ultimately a community project, because it involves a range of players, from individuals to the world, each making choices to reduce their impacts. A single person will have a pretty limited impact, but communities can change the world. And you need to have a true partnership with the communities you are working with, in order to actually make real change. This highlights the student's belief that sustainability is really a cooperative project. In general, students had a positive experience in the course and with its CBL aspect. The figure presents a word cloud of all of the reflections from the 2018 course offering. The words presented in the largest font were mentioned the most often in the student reflections, which include: sustainability, community, learning, and students. This brief reflection, employed during the 2018 course offering, is relatively simple to analyze, based on utilizing the frequency of words in the responses. It was applied to relatively short responses and represents one opportunity for student reflection. The major issue with this technique is that it only analyzes the frequency of words or short phrases, and not the context in which they are used, which would allow for the identification of themes within the responses themselves.

Emergent themes are identified through the usage of content analysis for the guided reflections from the 2020 offering of the course. Particular themes were identified in the majority of the responses, with the most common being that CBL provided a genuine context for learning about sustainability. Eighty eight percent of the students described this, and in particular, as a way to make sustainability real to them and to clarify their learning on somewhat abstract topics. The second most commonly identified theme was the notion that the CBL component of the course provided a deeper understanding of the material taught in the course (67%). Several students described that they worked diligently on this project because it would go to a real community partner, and was not only for a grade. Two intertwined themes which were commonly identified in the student reflections was the idea that working with a real community illustrated the complexities and tradeoffs inherent with sustainability work, and a related theme of the barriers inherent in completing this type of work, identified by 58% and 47% of the students, respectively. Other themes emerged, but were largely in the minority, such as sustainability as paramount to well-being, that sustainability is inherently a community endeavor, and relating sustainability to the circular economy. Based on the prevalence of the themes presented in the students' reflections, is it reasonable to conclude they in general found the inclusion of a CBL component in the sustainability engineering course to be beneficial to their learning.

Conclusions: IE is the "science of sustainability" and its application. Incorporating a CBL element into an IE course over several offerings has provided a positive experience for the students in the course, while providing them an opportunity to experience sustainability in action. The level of knowledge gained by the students is comparable to the first PBL offering of the course, and across the multiple years of offerings. Operationally, providing an experience like this for students is not all that different from the PBL methods which are commonly utilized in engineering courses, to teach topics such as IE and LCA, although CBL is more time and resource intensive. This combination of methods allows for IE education to be extending from working with industrial partners to improving the communities which the students live and work in. Multiple methods of assessment of learning ranging from the knowledge survey to the analysis of guided reflection are presented along with the tradeoffs comparing ease of use with insight generated.

sustainability
community
learning
based
students
sustainable
can
also
problems
real
learn
world
work
social
projects
project
teaching
communities
experience
knowledge
impacts
provides
classroom
three
different
improve
CBC
allows
partners
helping
paradigms
solve
apply
faced
helps
real-life
making
help
concepts
understand
problem
environmental
teaches
shows
important
economic
personal
background
setting
role
academic
challenges
practice
approach
learned
opportunity
places
life
skills
gave
gives
benefit
needs
better
Students
just
need
use
isn't
long-term
environment
class
must
welcome
teach

121d Chemical Engineering of Coffee: A Course Prioritizing Experiential Learning (and fun)

William McCaffrey, Vinay Prasad

Chemical Engineering is at a cross roads. Our traditional applications such as energy and polymers no longer capture the imagination of many young students. Ironically, the core concepts of a chemical engineering education such as heat and mass transfer, separations, thermodynamics, fluid mechanics, reaction engineering and systems engineering are all vastly important underpinnings of the transitions that must occur to support our future economy. Taking advantage of the statistic that Canada is one of the top 10 countries in the world for coffee consumption per capita, with an average of 6.5 kg of coffee consumed per person per year, we have been using coffee as a way to introduce chemical engineering to the general public. Coffee roasting and brewing involves nearly every aspect of the chemical engineering curriculum. An undergraduate club successfully modified our espresso machine to have PID control. The response has been so strong and positive that we have launched a senior elective course on the Chemical Engineering of Coffee that builds on foundational courses that the students have already completed.

The Chemical Engineering of Coffee course uses an experiential learning approach to explore chemical engineering concepts to all aspects of the coffee process, from bean to cup. We have embraced the experiential learning approach for this course emphasizing the four stages of feeling, watching, thinking and especially doing. The course is split between the class room and laboratory, with equal time for lectures and laboratory sessions and equal weightage for laboratory reports and exams in the grading. We explore many chemical engineering aspects of coffee making, including the impact of reactions during the roasting process, mass transfer and flux during brewing, the effect of water chemistry, pressure driven flow though porous media and scale up. Energy use and conservation during the roasting and brewing processes is also be investigated. In at least a minor way, the course helps counteract the perception that chemical engineering designs and operations are difficult to reduce to practice at small scales, since a unit operations perspective can be used to explain each of the aspects mentioned above, and also related to larger scale industrial operations.

Learning that is considered experiential involves reflection, critical analysis and synthesis; opportunities for students to take initiative, make decisions, and to be accountable for the results; opportunities for students to engage intellectually, creatively, emotionally, socially or physically; and a designed learning experience that includes the possibility to learn from natural consequences, mistakes and successes. Kolb's cycle of experiential learning [1] emphasizes concrete experience, reflective observation, abstract conceptualization and active experimentation. In our course, the laboratory component embodies the concrete experience (and being able to smell and drink your laboratory results doesn't hurt, either), and reflective observation is enabled by making the connections to traditional chemical engineering. Abstract conceptualization and active experimentation are emphasized in the second half of the course, where the labs are significantly more open-ended. For example, the students are presented with Darcy's law and asked to design and conduct experiments to test the validity of the law for pressure driven coffee brewing using a manual lever-type espresso device. They have to identify the dependent and independent variables, collect data while varying one of the independent variables and then, of course, taste the results.

In addition to being an excellent instructional tool, especially in the experiential learning context, the course is proving to be invaluable in aiding our outreach and recruitment efforts. Coffee, and making it, is ubiquitous and relatable, and a simple statement to the effect that everything about coffee is related to chemical engineering immediately awakens interest in the basis for that statement, which leads to curiosity about chemical engineering. In this context, we aim to also launch a non-mathematical short course for the general public in the near future.

References:

1. Kolb, David A. Experiential learning: Experience as the source of learning and development. FT

press, 2014.

RANCILIO

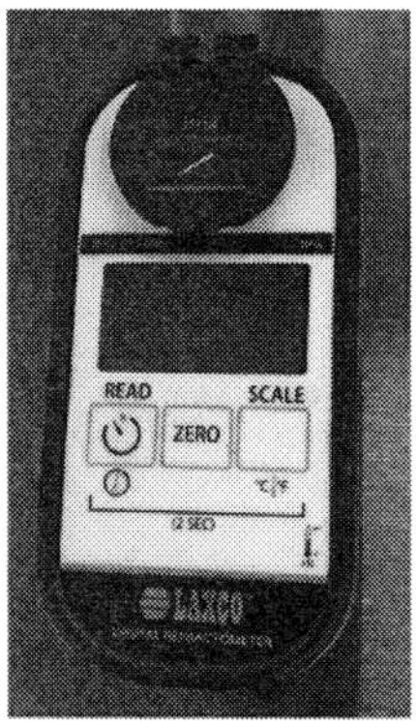
READ
SCALE
ZERO
LAXCO

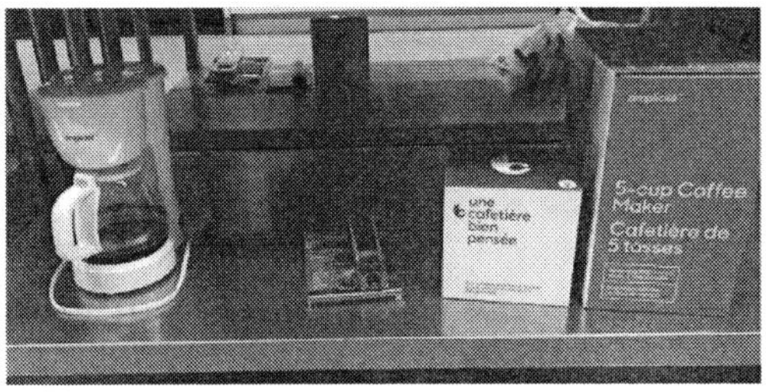
une cafetière bien pensée
5-cup Coffee Maker
Cafetière de 5 tasses

ACTIVE AND ADAPTIVE LEARNING UTILIZING A PROJECT-BASED APPROACH

Christine Leah Abalo and Abbas Ghassemi, PhD
University of California, Merced, Merced, California, USA

Biography

Abbas Ghassemi

Dr. Ghassemi, Professor Emeritus of Chemical Engineering, has been a Teaching Professor in Civil and Environmental Engineering at the University of California Merced since February 2018. Prior to that, Dr. Ghassemi served as Director of the Institute for Energy and Environment as well as Executive Director of WERC, a Consortium for Environmental Education and Technology Development, and Professor of Chemical Engineering at New Mexico State University from 1999 to 2015. Professor Ghassemi has extensive expertise in the areas of innovative programs in education, research, outreach in renewable energy, advanced water treatment, carbon cycle, carbon generation and management, air pollution control, and biofuels. As a teaching professor, he co-leads an experiential learning program that includes senior-level capstone and modeling and design courses. Additionally, Professor Ghassemi teaches classes in heat transfer, renewable energy fundamentals, and environmental engineering fundamentals.

Christine Leah Abalo

Christine received her BS in Chemical Engineering in the Philippines. She is currently a second-year PhD student at the University of California, Merced. Her research concerns nutrients and harmful algal bloom management in the Sacramento-San Joaquin River Delta in California. Christine is also a Teaching Assistant for two Civil and Environmental Engineering Department classes; one is about the environment in crisis for underclassmen, and one is teaching Professor Ghassemi's class, which focuses on designing energy systems.

Abstract

In modern academic settings, various learning approaches aim to create an inclusive and engaging environment. One of the most effective among these is active and adaptive learning within a project-based learning (PBL) framework, which immerses students in complex, real-world problem-solving. Research suggests that adaptive learning promotes higher retention, faster graduation rates, enhanced job placement, and greater professional advancement, particularly for first-generation and underrepresented minority students who may lack external resources. Evidence shows that an adaptive approach can narrow significant educational equity gaps in STEM fields, especially for Hispanic and other underrepresented students in California's Central Valley. This paper evaluates the outcomes of implementing a PBL approach in a junior/senior-level Environmental Engineering course at the University of California, Merced. The course challenged students to model and design an off-grid, low-energy-cost, sustainable power system for a cold storage facility. Guided by faculty facilitators, students worked in teams to apply scientific principles and develop innovative, quantitative frameworks for evaluating energy systems. This paper discusses the project outcomes, showcases student designs, and offers insights into lessons learned.

Introduction

Developing students' critical thinking to tackle real-world challenges is a core objective in higher education (Azid & Kiong, 2022). New teaching methods like adaptive learning are particularly impactful in STEM fields, as they adjust educational content based on students' needs and preferences (Bernard et al., 2019). Unlike traditional teacher-centered methods, adaptive learning emphasizes inquiry, iteration, and interactivity, fostering unique, individualized learning experiences (Hattie, 2008; Zhang, 2023). As a result, adaptive learning provides a unique learning experience to students by accounting for their individual differences.

A compelling way to implement adaptive learning is through project-based learning (PBL), where students address real-life problems collaboratively, enhancing communication and creative problem-solving skills (Dziuban et al., 2018). Through the PBL approach, students not only learn concepts of fundamental disciplinary subjects in the context of real-life scenarios they face daily, but also the value of teamwork and collaboration in solving the problem. Since its early application in medical education in the 1960s, PBL has proven to improve learning retention and facilitate interdisciplinary learning (Hidayatullah et al., 2020). However, the approach's relative merits and challenges compared to traditional methods continue to be explored (Martin et al., 2020).

This paper presents an analysis of a PBL-centered engineering course at the University of California, Merced. The students, primarily from underrepresented backgrounds in California's Central Valley, tackled the design and modeling of sustainable energy systems. This approach provided them with practical exposure to energy systems and teamwork while enhancing their adaptive learning experience.

Methods

This section presents the PBL method applied in the course. Four key PBL features were implemented in class: (1) having a facilitator who is an expert on the problem; (2) using open-ended questions to stimulate students' learning; (3) assigning teams to work on the problem collaboratively; and (4) assessing students' learning.

Professor And Teaching Assistant As Facilitators

In the PBL approach, educators serve as facilitators, providing strategic guidance rather than direct instruction. Effective facilitators are crucial for student success, as demonstrated in a meta-analysis linking facilitator experience with improved learning outcomes (Leary et al.,

2013; Arani et al., 2023). In this course, facilitators with expertise in adaptive learning created a supportive environment for students to explore complex problems. Although this approach is unique and offers new experiences to those students who were only exposed to traditional learning structures, the role of the facilitator is viewed as the most critical component for improving students' learning (Leary et al., 2013, Arani et al., 2023). For example, a meta-analysis explored the correlation between the facilitators' learning strategies and students' learning outcomes. The results showed that those educators who are well-trained and experienced in implementing the PBL approach in class have significant relationships with students' achievement. In comparison, untrained educators did not achieve the intended result of the PBL approach, and the students' learning outcomes were similar to those of educators who used a traditional learning structure (Leary et al., 2013).

Use Of Problems To Stimulate Learning

Students designed an off-grid, cost-effective, and sustainable energy system for a cold storage facility in California's Central Valley. The open-ended problem engaged students with practical engineering challenges and required them to synthesize knowledge across disciplines. Facilitators refrained from giving specific hints, encouraging students to develop flexible problem-solving skills and use the integrated experience of design and basic scientific principles in the design of the necessary components, processes, and systems.

Learning In Small Groups

PBL emphasizes team collaboration. Students were assigned to diverse teams, ensuring a mix of environmental, civil, and mechanical engineering backgrounds, and with balanced gender representation. Each team selected a leader to coordinate tasks, allowing members to contribute specialized knowledge and work cohesively toward a solution. Working in a small group structure helps the students to equally distribute the workload among team members and allows the whole team to tackle the problem that would have been too difficult to solve alone. Throughout the course, facilitators ensured consistent engagement within each group throughout the project.

Appropriate Assessment

Assessment involved both peer and facilitator evaluations. Each team member assessed their performance and effort, identifying strengths and areas for improvement. Facilitators graded teams based on comprehensive reports and presentations, focusing on problem comprehension, content quality, and solution viability.

Results and Discussion

Student's Approach to Problem Solving

The students worked collaboratively in teams to address the energy needed to operate a specified refrigerated cold storage facility. Each team proposed varying quantities of photovoltaic panels, inverters, charge controllers, and battery systems needed for an off-grid power system. Facilitators provided minimal guidance, prompting students to utilize relevant engineering principles independently. Iterative team discussions fostered adaptive problem-solving, as students refined their approaches. All teams were free to use any software, AI tools, and other resources that might be helpful in answering the problem. Through interactive discussions among the teams throughout the semester, the students continued refining their approaches while utilizing various specific principles to arrive at their findings.

Performance Assessment

Students were evaluated based on a 12-page technical report and an oral presentation. Average scores were 325 for reports and 89 for presentations. Figure 1 shows the variability of team scores. Several factors could explain the variation in team scores such as time management, adherence to formatting criteria, and report completeness. Common challenges included inconsistencies in calculations, incomplete economic and environmental analyses, and citation gaps. These issues affected some teams' credibility but highlighted areas for improvement in future PBL implementations.

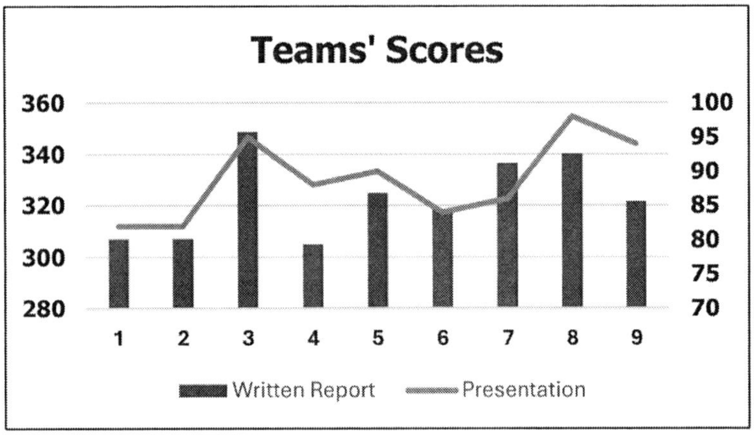

Figure 1. Assessment scores of each team.

Discussion

Each team came up with unique solutions to the open-minded problem. The facilitators only provided minimal guidance to the students. Students spent most of their class time engaged in team problem-solving, fostering deep critical thinking and collaboration. PBL's student-centered framework aligns well with active, adaptive learning principles, which emphasize knowledge construction through social interaction and self-direction (Yew & Goh, 2016). By fostering these skills, PBL prepared students for professional challenges that require interdisciplinary problem-solving and teamwork. One of the main expected outcomes of only scaffolding students' learning was to make the students think deeply about the problem as they make sense of their ideas, solutions, and opinions through discussion and sharing as a team. The reason why the PBL approach appeals to many educators in different fields is that it offers an instructional framework that supports active and adaptive learning, team discussions and learning, and iterative solutions (Frank et al., 2023). The framework of the PBL approach is premised on the belief that students learn effectively when they construct and co-construct ideas through social interactions and self-directed learning (Yew and Goh, 2016). Although the results of the assessment scores varied among teams, the intended result of the PBL approach was successfully achieved.

Conclusion

The PBL approach in this course equipped students with essential teamwork, critical thinking, and problem-solving skills. They gained experience in analyzing, designing, and evaluating energy systems and learned to overcome challenges through collaboration and adaptability. This approach not only helped students understand complex engineering concepts but also provided a practical foundation for addressing real-world problems within professional standards and constraints. To maintain the success of implementing the PBL approach in the course, the facilitators should continue to be well-trained and well-informed in the PBL process.

References

1. Azid, N.; Kiong, T.T. Cultivating thinking skills in education. *Springer eBooks*, **2022**, 355–377.

2. Herodotou, C.; Sharples, M.; Gaved, M.; Kukulska-Hulme, A.; Rienties, B.; Scanlon, E.; Whitelock, D. Innovative pedagogies of the future: An evidence-based selection. *Front. Educ.* **2019**, *113* (4), 1–14.

3. Bernard, R.M.; Borokhovski, E.; Schmid, R.F.; Waddington, D.I.; Pickup, D.I. Twenty-first century adaptive teaching and individualized learning operationalized as specific blends of student-centered instructional events: A systematic review and meta-analysis. *Campbell Syst. Rev.* **2019**, *15* (1–2), 1–35.

4. Hattie, J. Visible learning: A synthesis of over 800 meta-analyses relating to achievement. Routledge, **2009.**

5. Zhang, L.; Ma, Y. A study of the impact of project-based learning on student learning effects: A meta-analysis study. *Front. Psychol.* **2023**, *14* (1), 1–14.

6. Dziuban, C.; Howlin, C.; Moskal, P.; Johnson, C.; Parker, L.; Campbell, M. Adaptive learning: A stabilizing influence across disciplines and universities. *Online Learning* **2018**, *22* (3), 1–39.

7. Hidayatullah, R.S.; Ariyanto, S.R.; Mubarok, M.H.; Yohannes, A. Collaborative problem-based learning: An analysis of problem-solving skills in vocational schools. *Int. J. Educ. Res.* **2020**, *1* (3), 209–217.

8. Martin, F.; Chen, Y.; Moore, R.L.; Westine, C.D. Systematic review of adaptive learning research designs, context, strategies, and technologies from 2009 to 2018. *Educ. Technol. Res. Dev.* **2020**, *68* (4), 1903–1929.

9. Frank, M.; Lavy, I.; Elata, D. Implementing the project-based learning approach in an academic engineering course. *Int. J. Technol. Des. Educ.* **2003**, *13* (3), 273–288.

10. Leary, H.; Walker, A.; Shelton, B. E.; Fitt, M.H. Exploring the relationships between tutor background, tutor training, and student learning: A problem-based learning meta-analysis. *Inter. J. Prob-Based. Learn.* **2013**, *7* (1), 1–66.

11. Arani, S.A.Z.; Sarani, A. Problem-based language learning: Why aren't teachers using it? *Social Sciences & Humanities Open.* **2023**, *8* (1), 100668–100668.

12. Yew, E.H.J.; Goh, K. Problem-based learning: An overview of its process and impact on learning. *Health Prof. Educ.* **2016**, *2* (2), 75–79.

121f Shaping the Engineering Identity: New Elements for a Tissue Engineering Elective

Andrea Vernengo, Justin Major

Research focusing on mental health among college students has emerged as a vital area of research. Given the substantial rates of college enrollment and the prevalence of mental health challenges within this age group (ACHA 2013), it presents an opportune arena for exploration. It is recognized that mental health is intertwined with one's sense of identity. Identity, in this context, encompasses self-concept, perceptions of one's place in the world, and aspirations for the future. A favorable perceived identity promotes overall well-being, while a negative association can lead to risky behaviors, difficulty in coping with stress (Samuolis 2015), and potentially lower graduation rates among college-aged individuals, including those in engineering disciplines (Hughes 2019). The overarching objective of this endeavor is to leverage the engineering classroom environment to equip students with effective tools for cultivating mental health and alignment with a positive identity, thereby enhancing success in college and beyond.

To achieve our overarching goal, we opted to utilize a senior/graduate level elective course named Tissue Engineering. Within this course, we implemented an integrated design project that tasked students with designing a medical device and simulating its commercialization process. In addition to the anticipated scientific and business skill-building aspects typical of a project such as this, we incorporated additional instruction and homework assignments aimed at familiarizing students with tools well-accepted to enhance mental well-being. To this end, four mini-lectures were delivered, focusing on explaining methods of practicing gratitude, mindfulness, and finding belonging, meaning and purpose. Concurrently, students were assigned four writing tasks aimed at bridging these mental health tools with entrepreneurship. These tasks encouraged students to journal from both their current perspective and that of a thriving entrepreneur. For instance, prompts for envisioning success included questions like, 'Who or what do you appreciate for your achievements?' and 'Picture yourself as a patient benefiting from the technology you've developed and marketed. Describe the impact on the patient.'' And, ''How do you effectively deal with stress in your professional role?'

The integration of activities into the project was seamless, requiring minimal additional classroom time. We are currently evaluating the impact of these activities through pre- and post-introduction surveys, which assess students' attitudes toward their identity within the engineering community, self-perceived competence, interest, self-recognition, recognition by others, confidence, and enthusiasm in their field, among other factors. Additionally, we aim to offer a flexible and adaptable model for other engineering courses, particularly electives, where the subject matter coverage can vary. It is conceivable that these integrated activities will enhance performance on any classroom design project and students' perceptions of their final products, particularly as storytelling has been identified as a means of identity transformation (Hull 2006, Larraguibel 2022). From a broader perspective, we hope this experience equips students with valuable wellness tools applicable not only in tissue engineering and entrepreneurship but also in various future endeavors.

References:

ACHA (2013). American College Health Association–National College Health Assessment II: Spring 2013 reference group executive summary. Hanover, MD, American College Health Association.

Hughes, B. E., Schell, W.J., Annand, E., Beigel, R., Kwapisz, M.B. and Tallman, B. (2019). 'Do I think I'm an engineer? Understanding the impact of engineering identity on retention.' American Society for Engineering Education.

Hull, G. A., Katz, M.L. (2006). 'Crafting an agentive self: Case studies of digital storytelling. .' Research in the Teaching of English **41**(1): 43-81.

Larraguibel, Y. S., Edwards, M., Vasseur, R. and Elola, I. (2022). 'Perceived changes of identity construction through digital storytelling:: A collective study abroad case study.' Porta Linguarum: revista internacional de didáctica de las lenguas extranjeras **4**: 193-214.

Samuolis, J., Barcellos, M., LaFlam, J., Belson, D. and Berard, J. (2015). 'Mental health issues and their relation to identity distress in college students.' Identity **15**(1): 66-73.

581a Teaching Labview across the Curriculum

Kristine Horvat

Data acquisition and controls software is frequently used by chemical engineers in both industry and academia. One such software is National Instruments^{TM} LabVIEW. To develop undergraduate students' skills with LabVIEW, students are frequently exposed to the software from the first through last year of study across several courses. In the first year, students are introduced to the program in a Chemical Engineering Technology course over three weeks. After learning the basics of the program, how to utilize LabVIEW hardware, and ways to design a useful user interface, students are asked to implement a data acquisition program for a thermistor after following a detailed guide. This LabVIEW unit culminates with an open-ended project where students need to design their own program to control the salinity of water in a tank. In the third year, students are reintroduced to the program and asked to build more complex LabVIEW programs for various Unit Operations laboratory set ups with less detailed guides. Finally, in the last year, students are asked to develop their own LabVIEW program for a Unit Operations experiment with minimal instructions and constraints provided. Overall, this has proven to be a useful way to challenge students to build their own LabVIEW skillset across program curriculum, and these skills can be adapted into other data and controls programs that students may encounter later in their careers.

581b Introducing Process Simulation Software Earlier in the Chemical Engineering Curriculum

LiLu Funkenbusch

Process simulation software, such as the Aspen suite and Unisim, are powerful tools that we can use to design multiphase, multicomponent, multi-unit processes. Typically, this software is introduced during the senior year in capstone process design courses. However, this approach often leaves students struggling to learn new software while also trying to design a complex process. Most process simulators require significant amounts of practice and students do not have the time in a single semester (or even two) to learn how to use them properly.

To help students ease into using process simulators, I have begun introducing the software earlier in the curriculum during the Separations course. While students do not yet understand how to design a full process, they are introduced to the basics of the software (how to add chemical components, choose a fluid package, etc.) and simulate basic separations processes. Many of these units could be solved by hand, especially early in the course. However, giving students low-stakes problems where they already know or could easily find the answer allows them to explore the software with less stress and pressure on being able to produce results.

In this talk, I'll discuss how I introduce the software and what types of problems I give. I will also present survey results outlining students' familiarity with the software before and after the course, as well as feedback from students who were taught Aspen during Separations and then proceeded to take the process design courses.

581c Beyond the Optimization in Shell and Tube Heat Transfer Exchanger

Carlos Mercado, Mario Andrés Noriega, Iván Gil Chaves

The applications of optimization tools in heat transfer equipment design are widely known and accepted by academia and industry. However, designs obtained by optimization tools cannot always be built or work with the efficiency predicted in the optimization models. In the present work, optimization through various methods was employed for a shell and tube heat exchanger equipment design, the optimal result was subsequently validated with simulation using the Aspen Exchanger Design & Rating (EDR V14) tool. The optimization offers an approximation of the shell and tube exchanger's correct design. Nevertheless, the optimization process does not include information about vibrational analysis or hydraulic details provided by the detailed simulation, and, in a lot of scenarios, the obtained optimal design did not pass the vibrational analysis and other design tests. Consequently, the optimization process in heat transfer equipment design must be reinforced with a final stage which includes the optimal design validation employing simulation tools such as the EDR V14 tool. This stage will allow the user to guarantee the real efficiency of the optimal design proposed.

581d Are ChE Student's Afraid of Programming in Python or Matlab? Why Not Use Polymathplus?

Robert Hesketh, Michael Elly

With the fourth industrial revolution well underway[1], the proportion of occupations requiring "high" or "medium" digital skills has never been greater. Among those most in demand are engineers skilled in computing and advanced problem solving to support the ongoing digitalization, networking, and automation. A natural place to teach this to our students is through an integration of numerical methods into every chemical engineering course. One method currently being used to do this is through the use of templates coded as python Jupyter Notebooks or MATLAB live Scripts. But many chemical engineering students chose chemical engineering because they didn't like programming. Other students took that required programming course, but these concepts were never used in any of their subsequent chemical engineering courses. Matlab and Python require programming expertise for the efficient use of variables, data types, flow control, functions, and debugging techniques. If these skills are not properly applied, it can result in time-consuming efforts, errors, and significant frustration. In contrast, PolymathPlus simplifies the process by abstracting many traditional programming concepts. It enables students to input numerical problems in a format that is both simple and intuitive, reducing the learning curve and streamlining the problem-solving process. This approach not only saves time but also allows students to focus more on the engineering principles at hand rather than getting bogged down by complex programming syntax and errors.

One of the reasons that numerical methods are not used is that traditional textbooks do not incorporate numerical methods. One textbook that has led the way in this area is Fogler's Elements of Chemical Reaction Engineering. In the text of this book he has all of his examples using POLYMATH and on his website[2] he presents the same examples in python, MATLAB, Wolfram CDF Player. Unfortunately even with all of these examples, professors still don't require students to use numerical methods.

One of the reasons that Professors don't use these methods is that they are too busy and are unable to devote time learning python or MATLAB. Others try using MATLAB or python, but get frustrated when the solution obtained by POLYMATH has a superior fit to an a known analytical solution. But all is not lost! A new version of POLYMATH is available for use called PolymathPlus. This new program is backwards compatible and can run all of the old polymath files (*.pol). In addition, a new feature of this program is that it can be used on nearly any device since it is now a web application. Yes Mac users can now use POLYMATH! So students can write down the equations needed to solve a problem and then type them into the program on nearly any web browser, press solve, and get a numerical solution. With the use of POLYMATH all chemical engineering students can solve problems requiring numerical solutions.

PolymathPlus has the capability of solving systems of linear or nonlinear equations, systems of differential equations, and linear and nonlinear regressions. A free, student and professional versions of PolymathPlus are available for use. This new version allows professors and students to solve chemical engineering problems that require numerical solutions without having to learn MATLAB or python.

[1] Klaus Schwab, "The Fourth Industrial Revolution: what it means, how to respond," https://www.weforum.org/agenda/2016/01/the-fourth-industrial-revolution-what-it-means-and-how-to-respond/ 1/14/2016 World Economic Forum

[2] S Fogler, "Elements of Chemical Reaction Engineering" https://websites.umich.edu/~elements/6e/01chap/live.html , 4/7/2024.

581e Experience with Efficient Solutions of Numerical Problems in Reaction Engineering Courses

Satish Parulekar

Experience in obtaining efficient solutions of problems in undergraduate and graduate reaction engineering courses using Mathcad and Matcont, a utility in Matlab, is discussed to improve understanding of fundamentals of reaction processes. The emphasis is on obtaining solutions with increased confidence and higher precision and with minimum effort by understanding what type of problem has been defined. Students are encouraged to not introduce unnecessary complications when attempting to solve a problem and solve problems using optimal effort. Example problems considered for illustration deal with solution of linear algebraic equations (kinetic parameter estimation and properties of small and large reaction schemes and metabolic networks), nonlinear algebraic equations (computations of equilibria for multiple reactions, isothermal and non-isothermal operations of well-mixed reactors with single and multiple reactions, and kinetic parameter estimation using nonlinear regression), and ordinary differential equations [ODEs] (isothermal, adiabatic, and non-adiabatic batch and steady state continuous flow tubular reactors), application of difference equations (population and system dynamics, stochastic processes), and symbolic algebra (rate expressions for catalytic reactions and characteristics of reaction networks). Illustrations involving ODEs pertain to initial value and boundary value problems. Analysis of multi-reaction networks is based on properties of atomic and stoichiometric matrices. A smart use of relations among atomic and molecular species and molecular species and reactions obviates the need for solving conservation equations for all species in the reaction network, which typically are systems of nonlinear algebraic equations and nonlinear ordinary and partial differential equations. Any independent reactions missing in the reaction network are identified in a systematic fashion. Reduced order reactor models for large reaction networks can be obtained by applying pseudo-steady state hypothesis for highly reactive species, such as free radicals and various forms of catalytically active sites, an approach also useful in flux balance analysis in networks of reactions in living cell cultures (metabolic engineering), extending to genome-scale metabolic networks. Symbolic algebra is very useful in this effort. Mathcad is interactive, with user-friendly graphics capabilities. Students are encouraged to display results using graphics for improved learning and understanding. The capabilities of Mathcad are of significant benefit in accelerating the learning and strengthening the fundamental knowledge base. The nonlinear reaction kinetics can lead to steady state multiplicity, single period and multi-period cyclic steady states, and chaotic attractors (nonlinear phenomena). The pathological behavior of reactors varies with variation in system parameters owing to these nonlinearities, systematic investigation of which can be facilitated by developing parameter continuation methods. This is greatly facilitated by using Matcont, a Matlab utility. Specific illustrations for nonlinear phenomena are provided using symbolic and numerical computations carried out using Mathcad and Matcont.

581f In-Class Smartphone-Enabled Viscometry Experiment Adaptable for Junior-Level Chemical Engineering Fluid Mechanics and Heat Transfer Courses

Ifeoluwa Babalola, Jose C. Contreras-Naranjo, Victor Ugaz

Laboratory courses are inherent to the chemical engineering curriculum. In laboratory courses, theoretical concepts are taught and reinforced. In addition, laboratory courses provide opportunities to strengthen professional skills such as teamwork, leadership, communication, experimentation, and data analysis. However, most chemical engineering curricula have laboratory courses in the senior year, while the corresponding theoretical concepts are taught in the sophomore and junior years. This creates an undesirable time lag that might limit the effectiveness of laboratory courses. Additionally, most laboratory courses use conventional experimental setups, which are expensive, lack portability, and require lots of space. Updating these experimental setups with recent technologies can be challenging and costly. Alternatively, laboratory activities can be incorporated into the chemical engineering curriculum before the senior year to better reinforce the students' professional and technical skills. Therefore, developing and incorporating in-class hands-on experiments adapted with novel, user-friendly, and portable technologies to teach theoretical concepts earlier than the senior year may be an effective way to reinforce professional and technical skills.

In this work, we embedded a classroom-safe viscometry experiment to teach capillary flow theory, rheological analysis from capillary filling dynamics, and heat transfer in junior-level fluid mechanics and heat transfer courses. The viscometry experiment, initially designed for a fluid mechanics course, was adapted for use in the heat transfer course. In both courses, groups were assigned based on their personality preferences. Using smartphones, students recorded the self-driven flow of a tiny sample droplet (< ~60 microliters) in glass capillary tubes of diameters 400 and 800 micrometers. Newtonian (water) and non-Newtonian (xanthan gum) liquids were employed as samples, and their rheology was probed at room temperature for the fluid mechanics course. Meanwhile, Newtonian water-glycerol solutions were examined at various temperatures for the heat transfer course. The smartphone videos were then analyzed using computational tools (MATLAB and EXCEL) to retrieve the capillary flow dynamics and characterize the rheology of the fluid. The findings from this research reinforce the advantages of hands-on and experiential learning in improving how students learn theory and professional skills such as teamwork and data analysis.

581g Enhancing Undergraduate Heat and Mass Transfer Education with Computational Tools

Donald Sebastian, Aycan Hacioglu, Mae Markowski

An integrated view of heat, mass and momentum transfer has been a foundational element of Chemical Engineering education since the publication of Bird, Stewart & Lightfoot's "Transport Phenomena" over sixty years ago. These topics share a common mathematical framework, that should facilitate knowledge transfer from one domain to the other. Furthermore, the visualization of chemical processes and equipment in terms of mechanisms for accumulation, advection, convection, conduction and generation is cardinal to quantitative treatment of the classic unit operations that have been a basis for chemical engineering pedagogy since the inception of the discipline.

There is no denying that Transport or its sub-disciplines lean heavily on classic mathematics to reduce generalized balance equations to useful outcomes. The mathematics required for entry-level problems should lie within the scope of an undergraduate engineering curriculum and students are expected to retain their foundational engineering math courses knowledge to apply analytical solution techniques to differential and partial differential equation techniques. In the heat and mass transfer courses, the key objectives are teaching students the transport phenomena concepts so that they can translate "word descriptions" of real systems to "mathematical models" and able to interpret these models.

Many curricula have separated fluid dynamics from heat and mass transfer to allow more curricular time for each. Nevertheless, the time budget of a semester-long course does seem adequate to expand on both problems set-up and on the mathematical steps used to reduce initial equations to a working result. Even less time is afforded to interpreting and generalizing on results. Too often, the lifelong skills we wish students to develop take a back seat to mechanics of solution that very few will use after graduation.

This work describes the total integration of MATLAB, Symbolic Math Toolbox and Partial Differential Equation Toolbox as a vehicle for problem solving in lecture notes, homework assignments, exams, and design projects for the Heat & Mass Transfer course at NJIT. Before the Fall 2023 offering, all lecture slides were recreated from "pen and ink" solutions to an equivalent form solved using MATLAB's Symbolic Math Toolbox. All lecture scripts were posted to the learning management system and homework assignments were heavily templated to coach proper technique by example. Exams were given as MATLAB Live Scripts, and although students could revert to paper and pen solutions, everyone in a class of 35 opted to work in the scripting environment. Similarly, open ended design projects – one each in heat transfer and mass transfer – were created in the Live Script environment and students wrote their project reports in that form as well.

The approach was piloted in Fall 2023 and has been repeated with the Spring 2024 cohort and will be repeated in Fall 2024. The results have been strongly encouraging. From an instructional viewpoint, the transformation permitted changes in depth and breadth. The fraction of class time devoted to problem definition, road-mapping the solution, and analyzing the results increased by as much as 25% over previous iterations of the course. In the past, due to time pressure mass transfer took a back seat to heat transfer – as little as 3-4 lectures in a 14-week semester. The increased efficiency afforded by the MATLAB-based approach freed up enough time to expand mass transfer to 8 lectures and an ability to dive into that topic with fewer preliminaries and greater use of an analogy between similar problems in heat and mass transfer. The approach also affords a seamless migration to more complicated problems, such as transient, multi-dimensional problems that can be solved with

MATLAB's Partial Differential Equation Toolbox. Without getting mired in the solutions steps, one can show that the relationships and dependencies derived from analytic solutions provide meaningful guidance to the behavior in more complex settings as well as validation of the quantitative application of simple models.

From a student performance perspective, the response and outcomes have been equally encouraging. After expected grumbling and resistance to change the students quickly took to the new format. The most notable outcome during the semester was the performance on exams. The quality of answers – even when incorrect – was demonstrably better. The coding environment encouraged more orderly approaches to problem-solving. Unit conversion and substitution errors were eliminated providing real-time feedback on the correctness of the solution. Answers with inconsistent units gave clear evidence that the solution was incorrect, and numerical values out of the bounds of reality also triggered reinspection of the solution.

Finally, the student response after the course has reaffirmed the value of the experience. They report using the techniques on their own to solve problems in subsequent courses like Reaction Kinetics and Separations. Several have reported a new-found interest in applied computing.

The opportunities to build on this foundation seem unlimited. MATLAB Grader, automated grading tool for MATLAB assignments, has the capacity to provide instant feedback and evaluation to increase the value of homework assignments. Integration of advanced equipment simulations is possible to better connect classroom models to practical application. MATLAB's inherent capability to handle data acquisition, process control, and advanced data analytics suggests an abundance of opportunities to enhance the core content with examples and applications relevant across the spectrum of process industries.

581i Integrating Advanced Process Control Topics into the Chemical Engineering Curriculum Using Simulink

Aycan Hacioglu, Harish Chintakunta, Ahmed Mekky, Jordan Olson, Melda Ulusoy

Typical undergraduate chemical engineering process control courses cover a wide range of topics, including dynamic models of processes, Laplace transforms, transfer function models, and the dynamic response of first, second, and higher-order systems. These courses also delve into feedback and feedforward control, as well as stability. Generally, students learn the theoretical aspects of process control and subsequently apply their knowledge in separate chemical engineering laboratory courses, projects, or their professional careers.

Given the growing academic and industrial interest in more advanced process control methods, such as model predictive control or reinforcement learning, instructors are seeking ways to integrate these topics into both graduate and undergraduate process control courses. To support instructors in incorporating advanced process control topics in a practical manner, we will provide Simulink-based teaching resources, documentation examples, and content that we have delivered in seminars or as guest lectures. Considering MATLAB and Simulink are extensively utilized computational tools in both industry and academia, these resources and activities are designed to familiarize students with the tools they will encounter in industry or academic research.

120a Problem Type, Location, and Difficulty of Auto-Graded Homework for a Material and Energy Balances Course

Matthew Liberatore, Samantha Yanosko, Grant Valentine

Online homework and interactive textbooks document students' work and generate big data for both students and faculty to interpret. Here, students' usage of and success on over 700 auto-graded questions within an interactive tool titled the Material and Energy Balances zyBook will be explored. Real-time auto-grading allows students to know when they have mastered a topic, while teaching assistants and faculty can see both individual and group progress reports without laborious grading, especially for larger classes. Our previous research examined reading participation and auto-graded problems at the course level, which will be reviewed in this presentation. Specifically, median reading participation over 93% for seven cohorts was measured, and median correct on auto-graded problems of 91% or higher for six cohorts was found.

Since auto-graded problems allow unlimited attempts, students can receive feedback and persist, which captures some of the key tenets of deliberate practice. Two recent cohorts' responses on hundreds of auto-graded questions will be examined with respect to problem type, location, and difficulty. Formative, single calculation problems with scaffolding appear in most sections, and more summative, multi-concept problems appeared at the end of each chapter (following the standard convention in chemical engineering textbooks). Problems requiring numerical answers be within a tolerance will be compared with multiple choice questions. New findings will show median percent correct was high (above 80%) for all problem types. Attempts before correct provides another valuable metric to distinguish between problem types with numeric problems taking more attempts than multiple choice. Finally, a metric combining both correct and attempts, called the deliberate practice score, provides a quantitative aggregate measure. This contribution will build upon a recent ASEE conference paper on this topic.

120b Program Effectiveness - Evaluating Student Learning of Key Concepts from Year to Year

S. Alireza Bagherzadeh, Jonathan Verrett

The {Department blinded} at the {University blinded} undertook a broad curriculum renewal of its core undergraduate programs from 2019 to 2022. During this time, programs it offers were holistically re-evaluated and major changes were put in place [1]. As part of this effort there was a re-evaluation of the methods of measuring program effectiveness. One outcome was the development of a tool for evaluating student learning of key concepts at various year levels [2]. This tool parallels developments at other institutions [3].

Quizzes consisting of conceptual multiple choice questions were posed to student based on concepts from core undergraduate courses. These questions aimed to be at a high-level and did not require sophisticated computation. Students were also given the option to select an "I don't know" answer. The purpose of the quizzes, namely evaluating one aspect of program effectiveness, was explained to students prior to taking the quizzes. Quiz completion was for a participation mark in one of their courses but the quiz grade itself was not used for evaluation. Students completed two quizzes at the start of each year, one corresponding to the prior year of study and one corresponding to concepts in courses to come (to which they may have not yet been introduced).

Data has been collected since the 2021-2022 academic year. Analyses have been performed evaluating student knowledge changes from year to year, as well as areas of challenge for class cohorts. This work will present a background on the quiz development as well as a highlight of the analysis results.

[1] {blinded for review}

[2] {blinded for review}

[3] P. Farand and J. R. Tavares, "A concept inventory for knowledge base evaluation and continuous curriculum improvement," *Education for Chemical Engineers*, vol. 21, pp. 33–39, Oct. 2017, doi: 10.1016/j.ece.2017.07.001.

120c Starting to Ungrade Chemical Engineering: Practical Considerations and Instructor Reflections

Christopher Chen, Allison Godwin, Luke Landherr, Adam Ekenseair

Ungrading is a broad assessment approach that works to intentionally focus students on learning by moving beyond traditional grading practices. Strategies that fall into this assessment approach include contract grading; self- or peer-assessment; mastery- or competency-based grading; and portfolio assessment. Students benefit from ungrading over traditional grading schemes by helping to reduce stress; providing flexibility in how they are able to demonstrate mastery of content and concepts; facilitating involvement in their own assessment; and increasing accessibility and equity in the course. Studies outside of chemical engineering also indicate these approaches can increase deep learning and retention. However, there are multiple factors in deciding to implement ungrading, the approaches that may be most effective in different contexts, and practical challenges to implementation, especially in chemical engineering, where there are few examples of ungrading from which to draw.

In this presentation, we aim to share a wide range of instructor experiences of using approaches that fall into the ungrading umbrella–from three schools and across the chemical engineering curriculum–to help address the few existing examples within the literature base in our field. Where previous presentations and papers have focused on the effectiveness of ungrading on student learning, our talk will complement previous work by instead considering the instructor's point of view. Instructor reflections and practical considerations for implementation in the classroom and chemical engineering curriculum will be discussed so that our peers in chemical engineering can bring this assessment approach into their own classrooms.

Examples of ungrading in chemical engineering that will be shared in this talk include:

- Mastery-based graded first-year introductory chemical engineering course focused on increasing student motivation at School A
- Achievement-based graded core Materials & Energy Balances and Thermodynamics courses at School B
- Ungraded core Transport and Controls courses at School C
- Competency-based graded chemical engineering core post-baccalaureate course for non-chemical engineering M.S. students at School D

120d Ungrading and Student Motivation: Allowing Room for Point Totals and Assessing the Impact on Student Productivity

Luke Landherr

Ungrading refocuses the emphasis of a course onto learning and the mastery that students should have achieved at the end of the course, as opposed to students focusing on achieving certain grades and defining their success on the grade. In general, grades are de-emphasized for greater focus on students learning, processing, and applying the instruction in the course, through means such as greater levels of feedback and discussion with each other and the instructor. Many approaches fall under the broader umbrella of ungrading, including contract grading, mastery-based grading, portfolios, and self-assessment, but the general approach allows an instructor to provide less stressful, more equitable assessment.

When flexibility is more centered at the core of the assessment approach, allowing students to adjust how they can prove their mastery, as well as the overall breadth, depth, and body of work of mastery to be showcased, changes can be made to the assessment system over time to provide better support for students of all backgrounds and enable their mastery development. In previous iterations of implementing ungrading, changes were made to deadlines based on student feedback to reduce the flexibility of the deadlines throughout the semester, which resulted in higher student productivity. Further suggestions for improvement included assigning a point total to each exercise to allow students to 'gamify' and better score their work as the semester went along. As this potentially introudced a larger change to how students might approach the course, it was worthwhile to evaluate what elements of the course design were connected to student motivation.

In the third iteration of implementing an ungrading portfolio approach into a process control course, students could self-determine the assignments they would complete and how many assignments to complete, with all student deliverables culminating in one portfolio of demonstrated work by the end of the semester, but now with a point total assigned to each exercise to further delineate the depth of mastery shown by work completed. To evaluate the continuing impact of the ungrading approach as a means of course design, student feedback was assessed using the MUSIC model for student motivation, evaluating factors including empowerment, usefulness, success, interest, and caring as relevant to student productivity and motivating for overall effort. Students had been more highly productive with the ungrading approach compared to a traditional homework-and-exam semester; however, adding in a point score component opened parallels closer to the more traditional approach. The results of the ungrading implementation over three years, student feedback, and analysis of student motivation will be discussed in this paper.

120e Adventures in Ungrading Material and Energy Balances

Christy West

The alternative assessment termed "ungrading" is a method in which achievement of learning outcomes is determined by some means other than quantitatively scored exams. Many faculty remain skeptical of such an approach, particularly for courses that serve as pre-requisites for others. This work provides examples of three iterations of implementing an ungrading mechanism in a material and energy balances course. The objectives of the ungrading experiment were to reinforce problem-solving skills through student reflection, to promote deeper and longer-term learning by eliminating the focus on scheduled exams, to improve trust and communication between students and the instructor, and to reduce student stress.

Student achievement of learning objectives was evaluated based on solving problems from sets that were distributed as we progressed through the course material. These were submitted weekly or biweekly, and the instructor met with each student to discuss their work and assess their mastery of the learning objectives. In the first course, without specific deadlines to master a topic, they still submitted a significant number of problems each week, and if they didn't feel they had reached mastery they would ask for additional problems to work so that they could continue to improve. The objectives were met, but the instructor saw opportunities to incorporate some "guardrails" to reinforce the expectations of professional behavior. In the second course offering, limits were set on final grade achievement based on timely submission of some additional materials including class preparation assignments, ethics discussion posts and SAChE modules. A student assistant was employed to give detailed feedback on submitted problems. Despite the additional structure added to the course design, this new group of students did not respond with the same level of responsibility as the first, and the learning achieved did not meet that of the first group.

Informed by the first two experiments, the third iteration of the ungrading experiment in material and energy balances built in an additional level of accountability for independent learning. Some deadlines for completing problems from chapter sets were established, still allowing enough time for flexible learning schedules. Additionally, comprehensive reflections on accomplishment of the course learning objectives covered in each chapter were required at the completion of each chapter, reinforcing the student responsibility for their achievement.

It should be noted that the class sizes in this study were too small for effective quantitative surveys. Thus, this work is largely a reflection of the instructor experience, with mostly anecdotal evidence of student achievement and perceptions. Still, as such, it serves as an example of continuous improvement in course design based on informed innovation, mindful observation, and reflective adjustment.

120f Chemical Engineering Curriculum Redesign: Motivation, Method, Findings, and Implications

Ifeoluwa Babalola, Samantha Shields, Debra Fowler, Victor Ugaz

The undergraduate chemical engineering program at Texas A&M has faced changes on multiple fronts during the past decade. These changes were made in response to institutional growth plans that led to a significant increase in enrollment, alterations to our course offerings and sequencing, the adoption of a joint first-year program within the College of Engineering, shifts in student demographics, and changes in student competencies and attitudes as recruitment pipelines were expanded. Amidst this evolving landscape, our faculty engaged in numerous discussions regarding our curriculum, student education and performance, and the competencies necessary for graduates to succeed. In Spring 2020, the chemical engineering department began an undergraduate curriculum redesign effort in collaboration with the Center for Teaching Excellence at Texas A&M, following their Program (Re)Design (PRD) model. The process involved collecting feedback from over 400 stakeholders, including current and former students, faculty, employers, and colleagues at other institutions, to determine the knowledge and skills our graduates should possess. We then used these data to develop a comprehensive list of 47 knowledge and skill attributes, each with four assessment points in the curriculum (Introductory, Adequacy, Proficiency, Mastery). The resulting 188-entry table comprises our curriculum competency rubrics, providing a framework to assess these competencies across multiple courses in our curriculum to ensure connectivity. We also gathered faculty feedback at multiple stages and made refinements based on the input. Finally, we used the competency rubrics to generate a course map where each knowledge or skill is linked to one or more courses in our curriculum. Additional notable results from the redesign process include the design of two new courses, increased recitation hours and intentional integration of preferred computational software into the curriculum. The implications of these findings in this work will assist other chemical engineering programs in their curriculum redesign process.

447a The Influence of Perceived Help-Seeking Outcomes on Student Access to Mental Health Support

Sarah Wilson, Sara Lamer, Ava Huth, Joseph H. Hammer

National data suggests engineering students are professional treatment for their mental health at lower rates than students from outside engineering. This can result in worsening of mental health conditions to more severe or chronic disorders. Therefore, this quantitative study aims to understand the mental health related help-seeking beliefs of first-year engineering students and their impact on students' intention to seek help for their mental health. Using the Integrated Behavioral Model (IBM) as a framework, a survey was conducted in December 2021 at a large public university with 452 first-year engineering students. In this study, we present the outcome beliefs that were predictive of students' intention to seek help for their mental health. In particular, outcome beliefs that were negatively correlated with intention to seek help were that seeking help would: 1) go against the expectations of the engineering community, 2) be a sign of weakness or an admission of defeat, and 3) result in poor treatment or discrimination from the mental health professional. Conversely, the outcome beliefs most strongly positively correlated with intention were that seeking help would: 1) help me feel supported, 2) help me improve my coping skills, 3) make me feel better, 4) help me find a solution to my problem, and 5) help me gain a better understanding of my mental health concern. These findings illuminate critical aspects of first-year engineering students' attitudes toward seeking mental health treatment and offer insights for developing targeted interventions to improve their help-seeking behaviors.

447b A Faculty Flowchart for Student Disengagement and Mental Health

Maureen Tang

Mental health problems have skyrocketed among young people and students, and professors are increasingly finding themselves in challenging student situations. Faculty training, if it exists, does not typically address non-cognitive problems. It is unrealistic and unfair to expect professors to serve as counselors or therapists, but as classroom instructors, professors have opportunities to identify troubling behavior and influence students towards constructive behavior and help-seeking.

We created a simple flowchart to guide faculty and TAs through challenging student situations from mental health crises to chronic disengagement. Answering questions such as 'are you concerned about behavior or disengagement?' or 'is it a sudden change or chronic? ' directs instructors to appropriate responses. Interactive links provide specific examples, resource links, and, very importantly, suggested language for addressing the student.

The flowchart was developed through interviews with department heads as well as support units including: Academic Advising, Campus Safety, Counseling Center, Student Health and Wellness, Student Life, Student Success. Panels of students and faculty were convened by the Undergraduate Student Government Association and Teaching and Learning Center, respectively, to provide feedback on preliminary iterations. Suggested language for responding to students was approved by experts in counseling and mental health as well as the Office of General Counsel. Altogether, approximately forty stakeholders including faculty, students, administrators, and staff contributed input and feedback. This tool is designed for a resource-lean environment, and accordingly requires no workshops, training modules, database management, or staffing. The flowchart is available for public download on the Teaching and Learning Center's website.

This talk will discuss lessons learned from the flowchart development and rollout process and their implications for future faculty development projects.

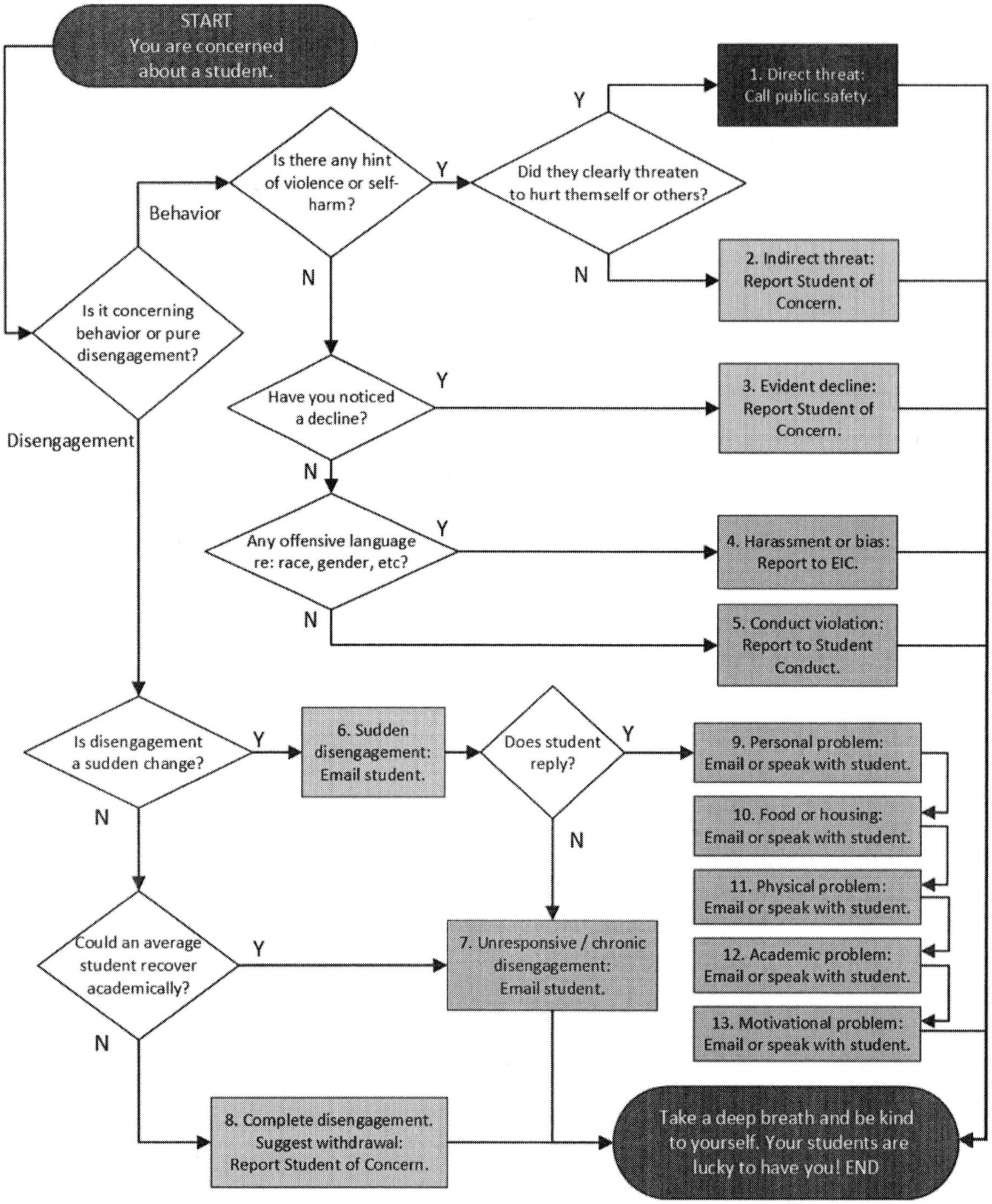
START
You are concerned about a student.
Is it concerning behavior or pure disengagement?
Behavior
Disengagement
Is there any hint of violence or self-harm?
Y
N
Did they clearly threaten to hurt themself or others?
Y
N
1. Direct threat: Call public safety.
2. Indirect threat: Report Student of Concern.
Have you noticed a decline?
Y
N
3. Evident decline: Report Student of Concern.
Any offensive language re: race, gender, etc?
Y
N
4. Harassment or bias: Report to EIC.
5. Conduct violation: Report to Student Conduct.
Is disengagement a sudden change?
Y
N
6. Sudden disengagement: Email student.
Does student reply?
Y
N
9. Personal problem: Email or speak with student.
10. Food or housing: Email or speak with student.
11. Physical problem: Email or speak with student.
12. Academic problem: Email or speak with student.
13. Motivational problem: Email or speak with student.
Could an average student recover academically?
Y
N
7. Unresponsive / chronic disengagement: Email student.
8. Complete disengagement. Suggest withdrawal: Report Student of Concern.
Take a deep breath and be kind to yourself. Your students are lucky to have you! END

447c Correlating College Student Participation in K-12 STEM Outreach with Sense of Belonging and Professional Development

Chris Torres, Julie Rorrer

University students, administration, and faculty implicitly understand the direct connection between high grades in formal coursework and graduation. The role of participation within informal community activities, however, such as K-12 science, technology, engineering, and math (STEM) outreach on student psyche and adjacent metrics for success remains unclear. In this study, surveys were distributed to STEM students via email and flyer canvassing at The University of Washington – Seattle, with 524 respondents from the Colleges of Engineering, Arts & Sciences, and the Environment over a 2-week period. Six prompts in Likert-type item style were scored from 1 (strongly disagree) to 5 (strongly agree) by respondents on self-perceived sense of belonging, qualification of employment in STEM fields, STEM fluency with children or non-STEM adults, and willingness to help and be helped with classwork. Rigorous statistical analyses with parametric and non-parametric techniques revealed significant differences ($p < 0.05$) between groups of students based on recent K-12 STEM outreach participation for all six survey prompts. For instance, students in the College of Engineering that volunteer in outreach express increased confidence in explaining STEM concepts to children compared to engineering students who have never volunteered in outreach. Interestingly, this difference in STEM fluency was absent amongst students in the College of the Environment and Arts & Sciences. Short answer responses revealed that students who never engage in K-12 STEM outreach are willing and excited to volunteer despite being unaware of available opportunities. These correlations elucidate benefits of K-12 outreach participation that foster positive student attitudes and provide motivation for academic institutions to advertise informal, community-based programming.

447d Academic and Career Advising to Maximize Student Degree Completion

Matthew Alexander

The academic advising of students is a critical component to their successful attainment of graduation with a BS degree in engineering, while engaging the student in career discussions can impact the student's success in attaining their first full time professional position. Our Hispanic-serving institution is a regional university with a long-standing and historic engineering program that serves to produce many engineers in our region of the state. However, as a regional, non-Tier 1 institution, with a 2.25 engineering course GPA requirement for graduation, our engineering academic departments must handle a wide range of academic capabilities in required engineering courses. Long gone are the days of freshman engineering or math instructors intoning the mantra to incoming freshman to "look to your left and to your right, and note that only one of you three will be sitting here come graduation time in four years". Retention of admitted students at all levels is a high priority in our institution.

Most recently, our College of Engineering utilized professional advisors to advise underclassmen students academically while they completed their STEM preparation courses, after which students "graduated" to advising with a faculty member for the rest of their undergraduate academic career. In just the past year, the college has changed this approach, so as to be in line with the rest of the university. In the new model, the students are advised by a professional advisor for the entirety of their undergraduate academic career, with select departmental faculty providing direction to the professional advisor for challenging cases. Select departmental faculty are also designated to provide career advising to the students at the student's request. In both approaches to advising, this faculty member has found that taking time for extended discussions and mapping out suitable course paths with a student that has challenges to completing their academic requirements to graduate usually results in the greatest success for the student. Retention trends within our department will be discussed as the measure of success with our college's new approaches to student advising.

447e Student Motivations for Enrolling in Online or Face-to-Face Sections of Material and Energy Balances

Peter McPartlan, Justin Shaffer

Chemical engineering departments may offer some sections of their courses in either traditional face-to-face format (F2F) or in online (OL) formats. Students therefore may have a choice in which type of section to enroll in and will make this choice for different reasons. The goal of this study was to assess the motivations behind student choice of enrollment in F2F vs OL sections of the materials and energy balance course (MEB). Students were given an online survey near the end of the semester which asked them about their motivation for their enrollment decision in addition to their thoughts on their interest in, the utility of, and their performance in MEB. Data will be summarized and presented by comparing F2F and OL student survey results in addition to student performance results in the F2F and OL sections. Recommendations will be made for chemical engineering programs to consider about the value of offering F2F and/or OL sections of core courses such as MEB.

447f Preparing Incoming Chemical Engineering Graduate Students from Varying Academic Backgrounds for Graduate Level Curriculum through Peer-Led Workshops

Sapna Ramesh, Willa Brenneis, Beth DiBiase, Jennifer Cole

This abstract describes a work-in-progress project.

Incoming graduate students to the Chemical and Biological Engineering department at [redacted] University come from a variety of different educational and personal backgrounds. Although several studies have assessed how race, ethnicity, and gender impact students' sense of belonging within STEM departments [1,2], educational background is understudied with regards to its impact on incoming graduate students' sense of confidence. Educational background can include factors such as differing undergraduate degrees (i.e., Chemistry vs Chemical Engineering), time off between undergraduate and graduate degrees, or exposure to US institutions of higher education vs international institutions. These factors may come into play during the first quarter of graduate school, where the lack of familiarity, feeling of being "othered," and feelings of being unprepared may lead to poor-performance in coursework and a negative impact on a student's sense of belonging in the field and department. [1,3]

Though many graduate students face similar academic challenges as undergraduates upon beginning a new program, fewer interventions have been introduced for this population. Support courses aimed at entering undergraduate to ensure that incoming students from differing backgrounds are all similarly well-prepared for undergraduate courses have found success in improving both confidence levels and academic performance. [4,5] One intervention aimed at graduate students, but with the goal of improving undergraduate education, included workshops to improve the ability of graduate teaching assistants, who then reported improved understanding of active learning principles post-workshop. [6] Most interventions geared toward graduate students target improving research skills, not coursework preparedness. [6,7] However, these courses have been successful in increasing students' confidence. [7]

In this presentation we will provide a review of the current literature on graduate student sense of belonging along with an analysis of some of the current interventions for improving confidence and sense of belonging for matriculating graduate students at US institutions. We will follow with the current state of our research proposal in the hopes of gaining some audience feedback during the presentation.

Our research is focused on two main goals: studying how an incoming Chemical and Biological Engineering student's educational background impacts their confidence coming into a graduate program, and how peer-led mini courses can help increase students' confidence and success in graduate school. In studying how students' educational background can impact their confidence going into graduate school, we will have valuable insight to take informed approaches to creating support programs for students to increase equity and success, in order to ensure that all students who matriculate in the graduate program have the resources to attain academic success in their coursework, regardless of their educational background. Lack of confidence and a sense of belonging upon entering graduate school can be attributed to a variety of factors including but not limited to personal identity and educational background. In addition to understanding the source of these challenges, we hope to ameliorate them by implementing peer-led workshops focused on reviewing common course material prior to the start of graduate courses.

(1) Stachl, C. N.; Baranger, A. M. Sense of Belonging within the Graduate Community of a Research-Focused STEM Department: Quantitative Assessment Using a Visual Narrative and Item Response Theory. *PLOS ONE* **2020**, *15* (5), e0233431. https://doi.org/10.1371/journal.pone.0233431.

(2) Clune, R.; Das, A.; Jasrasaria, D.; Rossomme, E.; Cohen, O.; Baranger, A. M. Development of a Week-Long Mathematics Intervention for Incoming Chemistry Graduate Students. *J. Chem. Educ.* **2023**, *100* (9), 3291–3301. https://doi.org/10.1021/acs.jchemed.2c00915.

(3) Pascale, A. B. "Co-Existing Lives": Understanding and Facilitating Graduate Student Sense of Belonging. *Journal of Student Affairs Research and Practice* **2018**, *55* (4), 399–411. https://doi.org/10.1080/19496591.2018.1474758.

(4) Raines, J. M. FirstSTEP: A Preliminary Review of the Effects of a Summer Bridge Program on Pre-College STEM Majors. *Journal of STEM Education: Innovations and Research* **2012**, *13* (1).

(5) Dockter, D.; Uvarov, C.; Guzman-Alvarez, A.; Molinaro, M. Improving Preparation and Persistence in Undergraduate STEM: Why an Online Summer Preparatory Chemistry Course Makes Sense. In *ACS Symposium Series*; Sörensen, P. M., Canelas, D. A., Eds.; American Chemical Society: Washington, DC, 2017; Vol. 1261, pp 7–33. https://doi.org/10.1021/bk-2017-1261.ch002.

(6) Roden, J. A.; Jakob, S.; Roehrig, C.; Brenner, T. J. Preparing Graduate Student Teaching Assistants in the Sciences: An Intensive Workshop Focused on Active Learning. *Biochemistry and Molecular Biology Education* **2018**, *46* (4), 318–326. https://doi.org/10.1002/bmb.21120.

(7) O'Clair, K. Preparing Graduate Students for Graduate-level Study and Research. *Reference Services Review* **2013**, *41*(2), 336–350. https://doi.org/10.1108/00907321311326255.

520a Teaching-Focused Faculty Workshop

Stephanie Velegol, Katie Cadwell

This workshop will include professional development topics that are specific to the needs of teaching-focused faculty in chemical engineering.

520b A 52-Week, Scaffolded Faculty Initiative Focused on Engineering Culture, Climate, and Education

Jennifer Fiegel

The faculty in chemical engineering at a predominantly white institution in the Midwest began a year-long journey in July 2022 to examine engineering culture and climate and their intersection primarily with engineering education. The goals of the project were to develop a department-wide learning process and to build a safe environment where faculty could consider and discuss challenging and thought-provoking content, have open dialogue, and build relational bonds. Discussions were aimed at bolstering our efforts toward building and nurturing a culture in which everyone is welcome, valued, respected, included, and supported. This work was facilitated by a faculty lead from the department that served as a Dean's Fellow within the college to support these efforts throughout the year. The presentation of this work will provide an overview of the scaffolded approach with examples of each element and the author's experiences in leading the initiative.

The scaffolded approach included three main elements: weekly challenges, guided workshops, and development of an action plan. Weekly challenge emails provide space for individual learning and reflection on a wide range of topics, including normalizing a culture of wellness, fostering a "weed in" culture, climate in the classroom, transparency in teaching, the power of words and stories, and student retention, persistence, and attrition. Hour-long guided workshops were run using the World Café method, where the power of conversation enables the group to share collective knowledge, question assumptions, and use our wisdom and creativity to confront difficult challenges in academia. Guided group discussions in the World Café workshops focused on examining relevant case studies, norms, and practices in engineering education. Small groups of faculty discussed ways we might be perpetuating the current culture and how we can interrupt or change our approaches to better support the success of all in our community. Faculty retreats focused on using the knowledge gained throughout the semester to develop an action plan to shape our future and make positive change happen. A DEI rubric developed by the PULSE network provided a framework for these discussions, enabling the group to baseline our current DEI efforts. Through this effort, the group identified areas of commonality where the department had already put in a lot of effort (the wins), areas where we need to collect data to support our observations or gain more information (questions or missing information), and key areas that we as a faculty wanted to dedicate time to (areas of improvement).

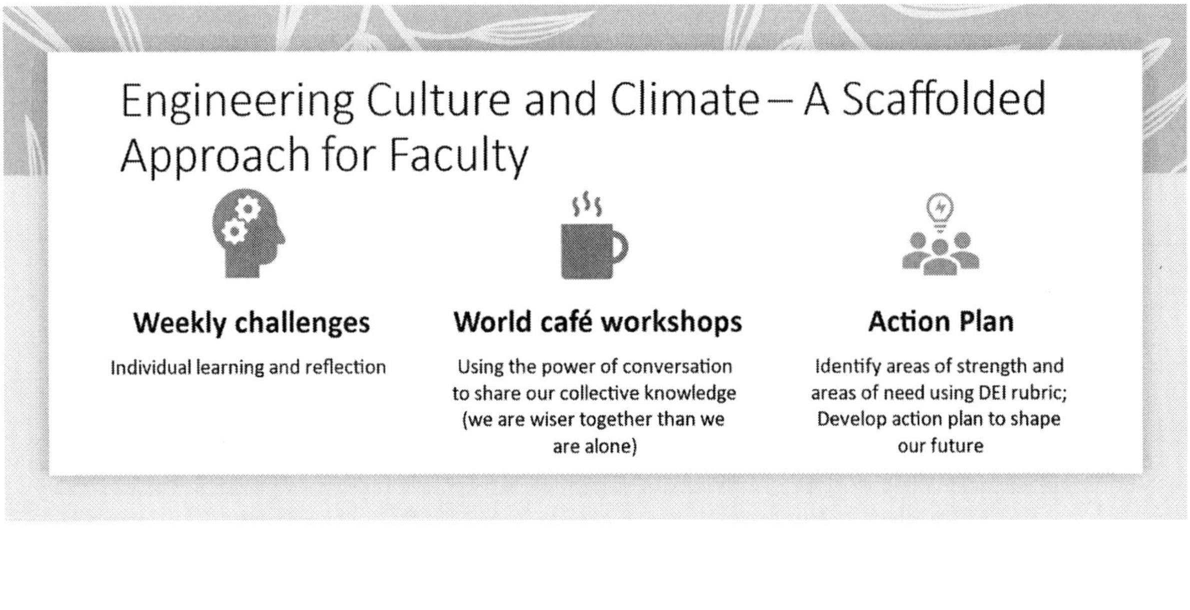

520c Salaries of Chemical Engineering Teaching Focused Faculty: How Are They Predicted?

Stephanie Velegol, Taryn Bayles, Lisa Bullard, Mechteld Hillsley, Katie Cadwell

In this presentation, we will share trends in salary data from a national survey for teaching-focused faculty, TFF, (non-tenure track) in chemical engineering. Previous work showed that the yearly salary ranges from $80,000 - $200,000. Surprisingly, the salary does not correlate with either years of teaching experience (ranging from 0 to 20 years) or with school ranking. The survey was sent out to 270 TFF in Fall 2023. There were approximately 100 respondents that represent various types of schools (R1, R2, PUI, etc.), sizes and ranks of program. Faculty were asked about a variety of factors including age, previous industrial experience, number of promotions, length of contract, job acquisition and job responsibilities. In this presentation, we will show if any of these variables correlate with salary. This will shed light on faculty remuneration and inform discussions of equitable compensation practices.

520d Evaluation of Teaching for Tenure Decisions

W. Vincent Wilding, Richard H. Swan, Tina M. Taylor

End-of-semester student ratings are the most commonly used metric for the evaluation of teaching for tenure decisions. However, these ratings, although valuable, are insufficient measures of effective teaching. We have developed a new methodology – *The Teaching Portfolio and Peer Review* – to provide a more substantial and informative evaluation of teaching to be used in the tenure process. This is being institutionalized across our university and is not only providing better evaluations of teaching in the tenure process but is also encouraging efforts to improve teaching and learning. This methodology may be useful for other academic institutions who are striving to elevate the standing of teaching in the tenure process.

254a Creating Virtual Reality Content for Core Materials and Chemical Engineering Courses

Keisha Antoine, Olufemi S. Duyilemi, Jaleel A. Thorpe

Virtual reality (VR) is a fully digital space in which a user is immersed in a virtual environment through the use of a headmounted display. The user's view of the physical environment is occluded by the headset, thus creating total visual immersion. VR also simulates other human senses (to a degree) like hearing, and grasping through the use of hand controllers. Many undergraduate chemical engineering students taking core chemistry and materials science courses struggle with spatial visualization of molecules, atomic arrangements, and crystallographic structure. The focus of this study is the creation of 3D holographic content that allows the student to be exposed to the 3D nature of chemical compounds. We utilized the VR environment of VictoryXR powered by ENGAGE to provide the immersive experience.

254b High Structure Course Design for Chemical Engineering Courses

Justin Shaffer

Calls for using evidence-based pedagogies have been expanding in the past decade so as to improve student learning and outcomes. High structure courses are designed to do just that as they prepare students to be actively engaged in the learning process via pre-class content acquisition and assessment, in-class active learning and problem solving, and after-class review. Significant literature has showed high structure courses in a variety of STEM disciplines to be efficacious over traditional didactic courses as students show increased performance, students feel more belonging, and achievement gaps are narrowed or closed. This talk will cover the basics of using high structure course design in chemical engineering courses including introductory thermodynamics and material and energy balances. Attendees will leave this talk with the knowledge, skills, and references necessary to design their high structure course from scratch or to modify an existing course into a high structure course.

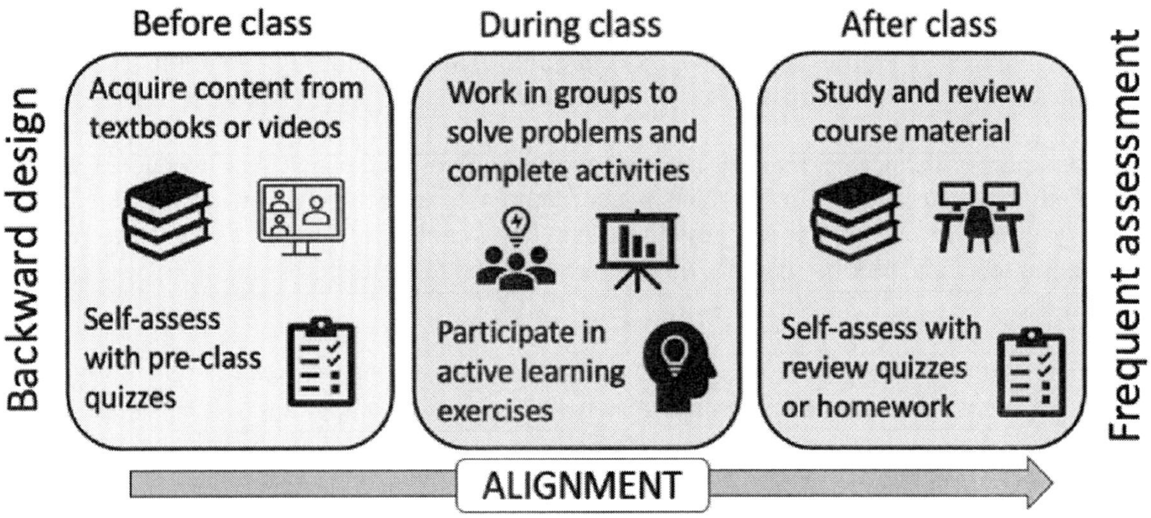

254c A Scaffold for a Project Assessing Communication, Reflection, and Self-Motivated Learning in Core Chemical Engineering Courses

Katelyn Dahlke

Developing projects for core chemical engineering classes is a challenge. Long technical questions are time consuming to develop and time consuming to grade. Another challenge in technical courses (outside of laboratory and design courses) is the inclusion and assessment of "softer" skills, including communication.

An end-of-semester project scaffold was developed for two different chemical engineering thermodynamic courses. Students select a problem that they struggled with (either from homework, exams, or quizzes) and redo it correctly. They must explicitly discuss what aspect of the problem they initially struggled with and what they have since learned. This portion of the assessment incorporates both self-motivated learning and a mastery learning approach. Students must also communicate this process to an expert in the field (the instructor) through writing and through notations in their technical work.

The second part of the project requires students to write a two-page paper for a lay audience related to the problem they selected in part one. This paper requires the student to constructively engage with a topic they found difficult, specifically by discussing it in simple terms for a non-expert to read. Compared to technical reports, grading time is minimal due to the requirement of layperson readability.

Future work with this project scaffold will investigate how to extend its use into other courses and robust assessment of student learning outcomes before and after completion of the project.

254d Preparation Assignments in a Flipped Classroom

Carl Lund

A reaction engineering course has been taught for many years using a flipped classroom approach. Prior to a typical class meeting, students are expected to read book sections and watch videos. The videos present information and illustrate the solution of problems in a format typical of didactic classroom lectures that were used before adopting the flipped classroom approach. At the start of the class meeting the information from the readings and videos is summarized and students are given the opportunity to ask questions. During most of the class meeting time, the students work to complete a problem similar to those presented in the readings and videos. Depending upon what information and problems have been considered in recent, preceding classes, the students may be guided through the solution to the problem. In all cases, the instructor circulates about the classroom observing student progress, answering questions, and correcting mistakes. If a number of students display a common misconception, the instructor interrupts the problem solving and presents a 'mini lecture, question, and answer session' dealing with that misconception. The students then resume solving the problem.

As the students work through the problem in class, the instructor repeatedly points out to them that getting stuck and now knowing what to do is a good thing (at this point in the course). They are told that each time they get stuck, it reveals a point where their understanding is incomplete or where they hold a misconception. In either case, as long as they ask for help (or compare to the instructor's solution when it is presented), they can fill in the missing understanding or correct their misconception. They are also reminded repeatedly that solving the problems themselves is fundamentally different from reading or watching as someone else solves the problem. The may understand the other person's solution completely and be able to explain it in detail, but still not be able to generate a solution themselves. The instructor repeatedly reminds them that the in-class problem solving enables them to learn how to use the information to solve problems much more efficiently than most can learn outside the classroom, the difference being the immediate availability of the instructor.

Students will benefit most from this flipped classroom approach if they indeed do complete the reading and watch the videos prior to the start of class. Unfortunately, over the years the instructor has observed that many students have not done the reading or watched the videos or that apparently the did so in a superficial manner. When they have not prepared effectively, they spend time reading or skimming through the pre-class information. Consequently, they progress through the problem slowly compared to better-prepared students. Instead of removing misconceptions or filling small gaps in their understanding, they spend class time gaining a level of understanding that they should have reached prior to the start of class. When other students ask questions or the instructor breaks into a 'mini lecture, question, and answer session,' they may not have sufficient understanding to even appreciate the point under consideration, let alone to learn from it.

Over the years, the instructor has tried in a few different ways to coerce the students to read the book sections and watch the videos prior to the start of class. Those efforts will be reviewed and assessed briefly, leading to the 'preparation assignments' that are currently being used. The preparation assignments are surveys that are administered through a learning management system. The is a preparation assignment due prior to the start of every flipped class. Students earn one point for each preparation assignment they complete. These points become a weighted fraction of their final course score, from which their grade is determined. The weighting of the preparation assignments is 10% of the course score, corresponding to approximately one letter grade. The effects of including the preparation assignments in the final course grade at this weighting level will be discussed.

The students are promised that the instructor will only see a summary of each preparation assignment during the semester. They are told that after the semester has ended and grades have been finalized, the instructor will be able to see individual preparation assignment responses. They are told that these responses will be used to improve the course and that their individual responses will not be shared with anyone. They are also told that the analysis of the data and the protocol for protecting them from being identified in any reports or research papers has been reviewed and approved by the university's IRB.

An example preparation assignment will be presented and discussed. Each preparation assignment begins with a multiple choice question asking how thoroughly they completed the readings and videos: not at all, skimmed, skimmed but with more thorough examination of certain topics, or thoroughly prepared. This initial questions was followed by a group of five point Likert-scale questions asking the level of their agreement with a statement. The statements were structured so that the students needed to reflect upon their preparation for the class meeting. Some of the statements focused upon their knowledge and others upon their ability to use that knowledge. They typically addressed a single bit of knowledge or a single ability. The things that these statements focused upon were the same things that the instructor would review at the start of class. The last question on each preparation assignment was an open response question: Enter any questions you have about the information for this class meeting. The instructor reviewed the last question just prior to class. Then to begin the question and answer period at the start of class, they were presented and discussed.

Several aspects of the preparation assignment responses will be considered. The response rate and apparent sincerity of the responses will be discussed. Correlations between response summaries and course performance will be examined. For example, for any one exam question, is there correlation between student reported level of completion of associated preparation assignments and are there correlations between Likert-scales on individual concepts and abilities and exam results. Finally, student self-awareness will be gauged by comparing individual responses to corresponding steps in exam problem solutions. An overall assessment of the effectiveness of the preparation assignments will be offered, based upon these aspects, and plans for modification will be presented.

254e Enhancing Learning Effectiveness in Chemical Engineering Via Connection to Real-World Engineering Applications

Stefanos (Steve) Nitodas

Engineering education for undergraduate students in the U.S. experiences several long-standing challenges, and there is a genuine concern from educational and industrial leaders that the U.S. may lose its competitive edge in science, technology, engineering, and math (STEM) disciplines.[1-3] A fundamental challenge in undergraduate engineering education lies in the disconnection between engineering disciplines taught in the classroom and the practical skills needed in the industry.[4,5] Even though engineering students build a solid foundation of the theoretical knowledge of their disciplines during their studies, they often lack opportunities to get acquainted with real-world applications of the engineering principles applied in companies and production plants. As a result, there is a serious mismatch between the contemporary undergraduate engineering education and what is required for the students to be successful as engineers in the modern industrial world.[5]

This work aimed at enhancing student learning engagement and comprehension through the utilization of scientific and technical videos in class, which can connect chemical engineering curricula to real-world engineering applications. These engineering curricula (courses) involved process control and nanotechnology engineering applications. In addition to the videos use in the classroom, peer engineers at local technologies companies visited the aforementioned chemical engineering classes as guest speakers and presented industrial case studies and technical challenges encountered and resolved in their production plants, aiming at improving the students understanding of the modern industry applications. Furthermore, the utilization of Piazza, a learning management system where students can ask questions in a forum-type format, promoted the sharing of opinions and knowledge between students on the topics addressed in class, and promoted exchange of insight about real-world problem solving.

The performance of the students was evaluated in this study and compared to that of students of prior years that did not use videos for the understanding of the pertinent engineering concepts and were not exposed to presentations from industry professionals. Targeted criteria were employed for the performance evaluation, including the ability of students to identify and solve complex engineering problems ability as well as to apply engineering design to produce solutions that meet specified needs with consideration of safety, social, environmental, and economic factors. The results demonstrated how the implemented instructional methods advanced students understanding of educational outcomes through active learning exercises related to the videos applications and the demonstrated industrial case studies. This work also showed how new learning media can benefit students with different learning capabilities who are marginalized by traditional learning modes, thus leading to greater persistence in STEM fields.

References

1. National Science Foundation, *"International STEM Talent is Crucial for a Robust U.S. Economy"*, National Science Board, **2022.**
2. Council, N. R., Committee on Prospering in the Global Economy of the 21st Century: An Agenda for American Science and Technology, Committee on Science, Engineering, and Public Policy. National Academy of Sciences: **2007.**
3. Sevo, R., Literature overview: The talent crisis in science and engineering. *Apply Research to*

Practice (ARP) Resources. **2009**.
4. Hubbard, S.; Lopp, D., A case study reflecting successful industry engagement: Utilizing industry projects as a mechanism for discovery and STEM education in aviation technology. *Journal of Education and Human Development* **2015,** *4* (2), 175-181.
5. Bucciarelli, L. L.; Kuhn, S., Engineering Education and Engineering Practice: Improving the fit. *Between Craft and Science: Technical Work in the United States* **2018,** *210*, 9781501720888-012.

254g The Care and Feeding of Industry Mentors for Senior Design

Lisa Bullard, Matthew Cooper

Over the last 23 years, the author has evolved from offering the same "cookbook" senior design problem to everyone in their 100+ person class, to soliciting and managing a portfolio of original projects each year such that each team has a unique design problem, most of which are industry-sponsored. This approach requires that the faculty member serve more as a project manager with a large stable of both industry and faculty mentors to manage, and the success of these mentor interactions has a large influence on the experience of both students and industry mentors during the capstone course. This presentation will discuss the mechanics of project recruiting and scoping to ensure that each project meets ABET expectations for capstone design. Project communication and selection will be detailed, and clearly stated expectations for both mentors and students will be shared. The development of deliverable descriptions and grading rubrics will be outlined. The trade-offs of having industry mentors pay for project sponsorship versus having them serve as technical mentors and graders will be discussed. The importance of consistent communication with mentors will be emphasized. Finally, the challenge of maintaining mentor relationships over many years and making the process sustainable will be explored.

198a Broadening Participation in Engineering: Professional Development for STEM Teachers in Pennsylvania Rural Counties

Taryn Bayles, Claudia Morrell, Sandra Staklis, Kevin Jordan

Over the last two years of this NSF BPE grant, the program team (engineering faculty and engineering students from rural high-schools, a nonprofit, research partner and advisory board) developed and refined a professional development (PD) program to guide rural-area secondary school teachers to integrate evidence-based engineering content, effective pedagogical practice and innovation indicators of education equity. The NEIR (Normalize, Empower, Inclusive, Relevant) model focusses on classroom equity integrated with the engineering design process used to connect science, technology and math content. Two PD workshops (ranging from 2-3½ days, supplemented by year-long Faculty Learning Communities) have trained twenty-five teachers from seventeen different schools (four counties), and eight returning teachers; in addition, six administrators from two different counties participated in various components of the PD program.

In terms of the three objectives for the research plan, the research study identified the following findings:

1. Indicators of successful model implementation emerged from teacher's feedback on how their classes changed through their use of the NEIR model. The changes observed by the teachers included:
    1. Projects that engage student in problem solving and the design process rather than kits or high structured activities
    2. Examples of how the teachers use NEIR in the classroom
    3. Classroom activities that required collaboration among all students, which included changes to classroom organization
    4. The teacher's role in the classroom changed from providing answers and instructions to serving as a facilitator and advisor, allowing students to work on challenges and failures on their own and with their peers.
2. Assessing the feasibility of implementing the curriculum in rural STEM classrooms: Teacher's feedback during learning community sessions, interview and focus group responses, and responses to the Stages of Concern (SOC) questionnaire from the Concerns-Based Adoption Model (CBAM) suggest that teachers were engaged with the program and found the model usable and feasible to implement.
3. Collecting initial data on the program's effects on the classroom environment and student's engagement and interest in engineering: the researchers collected evidence on changes to the classroom environment through feedback from teachers. All teachers indicated that use of NEIR model and the engineering design process increased student engagement and enthusiasm for coursework overall but acknowledged that engagement for some classes and students remains a challenge.

This presentation will describe the details of the PD, the NEIR model, engineering curriculum description, teacher and student survey results, and lessons learned from this innovative approach to equity in engineering education in secondary schools.

198b Virtual Reality Movies and Marshmallows: A Crystal Structure Workshop for Middle School Students

Sarah Stratton, Matthew Montemore

Crystal structures and their atomic builds can be a difficult topic for middle school students to conceptualize. To help students better understand these structures at the atomic scale, we have designed a workshop that uses movies and crafts to give students an introduction to crystal structure. This hour-long workshop starts with an interactive presentation on atoms and molecules that gives the background on crystal structure. The Visual Molecular Dynamics (VMD) program was used to create movies that give students a virtual-reality (VR) view of four different crystal structures – sapphire, DNA, salt, and sugar. Using provided VR glasses, students work in small groups or pairs to fill out a worksheet that uses the scientific method to make observations and sketch the crystal structures they viewed. Class-wide discussions occur between each structure observation to identify patterns within the crystal. The students then use their observations to create a model of one of the four crystal structures using marshmallows and toothpicks. This workshop is a simple introduction to complicated topics on the atomic scale. Students and teachers have reacted very positively to both the VR aspect and the hands-on building of the structures. We will share details of the materials of this workshop – VR glasses, phones, and building materials, along with how this workshop could be scaled up to a more advanced level.

198c Outreach at the Shepherd Chemical Company: Activities, Successes, and Motivations

John Stueve

The Shepherd Chemical Company is a 108-year-old specialty chemical manufacturer, based in Norwood, Ohio. Over the past ten years or so, Shepherd's outreach efforts have grown to encompass K-12 students, and we've further improved our retention efforts for co-op students. We how have recurring outreach events with several local elementary and high schools, have been encouraging employees to participate in local science fairs as judges, and have involved our engineering co-op students in these outreach activities, thereby continuing these outreach traditions in the next generation. Remarkably, there has been no general company directive or push from management to perform this work. Each of these efforts are run independently of one another and were initiated simply because the interested party had the right connections and a desire to share their enthusiasm for chemistry and engineering. This presentation will go into more details about these efforts and share the successes and growth we've seen in both our outreach programs and the students.

198d Cultivating Engineering Identity in K-12 Students through Engaging Activities: A Three-Week Summer Program

Betul Bilgin

The Department of Chemical Engineering has developed and executed a pioneering three-week outreach program aimed at inspiring future generations of engineers by fostering an engineering identity among students from 4th to 12th grade. This initiative involves a carefully curated series of activities designed to introduce young learners to the field of engineering, covering a broad spectrum of field of chemical engineering. Through hands-on projects, interactive workshops, and real-world problem-solving scenarios, the program seeks to demystify engineering concepts and showcase the impact of engineering solutions on society.

A critical component of this outreach effort is the implementation of surveys designed to measure shifts in engineering identity among participants. These surveys, administered at the beginning and end of the program, aim to assess changes in students' perceptions of engineering, their interest in pursuing engineering as a career, and their confidence in their abilities to engage in engineering tasks. The data collected offers invaluable insights into the effectiveness of different activities in enhancing students' understanding and enthusiasm for engineering.

This abstract presents an overview of the program's structure, from the planning and execution of the activities to the methodology and preliminary findings of the engineering identity surveys. By evaluating the impact of early engineering exposure on students' attitudes and aspirations, this program contributes to the broader goal of diversifying and strengthening the future engineering workforce. The insights gained from this initiative are expected to inform future efforts in K-12 engineering education, highlighting the importance of early, engaging, and inclusive outreach strategies.

198e Using Educational Comics in Outreach Efforts to Promote Interest in Chemical Engineering

Nethra Iyer, Ira Hysi, Luke Landherr

Opportunities and growth in larger STEM fields over the past few decades have driven the need for broadening participation and support in engineering education in order to provide students with the necessary skills and mindset to succeed in engineering jobs. A major factor in these efforts is the need not just to increase enrollment in engineering majors, but to ensure that enrolled students are retained within the field over the course of their studies and once they enter the workforce. Retention becomes an even greater concern to ensure participation by women and underrepresented minority (URM) students can be lifted towards a more equitable level as that of white male students. Some of the greatest concern in retention has focused on self-efficacy beliefs of engineering students. Studies have shown that the efficacy of STEM students is directly related to their interest in the field, persistence to remain within the field, and their resulting achievements. Students who leave the STEM fields are more likely to have lower efficacy beliefs, even if they exhibited similar grades to students remaining in STEM; this difference is also particularly notable for women students

An integrated STEAM (science, technology, engineering, arts, and mathematics) approach to engineering education has been a broad effort over the last few decades, with objectives ranging from driving interest in STEM in young learners, to enabling better classroom instructional approaches, to creating more interdisciplinary collaborations in learners and instructors. Many different efforts have been utilized to produce STEAM education in different fields, including video series, in-class modules, and broader creative projects. Several studies have shown that these STEAM efforts have had positive impacts on student interest, understanding, and self-efficacy, and may lead to greater retention. One STEAM approach that has been shown to be effective in a wide range of STEM fields is through the use of educational comics, combining text with visual presentation to enable greater conection between the reader and the content, while also providing a means of implementing greater representation in the educational tool through diverse depiction of the characters in the comics. Utilizing comics in chemical engineering undergraduate courses has been demonstrated to increase student interest and confidence with potential further benefit to understanding; similar potential exists to utilize such an approach to promote interest and self-efficacy in younger students in chemical engineering concepts.

In order to help bridge students from K-12 education and in general first-year engineering programs into chemical engineering, a series of educational comics were developed in collaboration with student and professional artists for outreach and formal learning purposes. These comics included a series of one-page comics connecting everyday applications like ice cream, hockey, and nail polish to chemical engineering concepts, as well as a longer 16-page comic highlighting the breadth of the different fields that chemical engineering encompasses, including medicine, materials, renewable energy, and food science. These comics were disseminated with K-12 andd first-year engineering students, with student understanding of chemical engineering as well as interest and confidence in the subject matter being assessed through a series of surveys. Self-efficacy and motivation of learners was studied with additional consideration of student learning styles. This paper will discuss the efforts in utilizing these comics with different groups of students and the results from their implementation.

198f Art Meets Science: Introducing K-12 Students to Chemical Engineering Concepts through Art

Betul Bilgin

In an innovative outreach initiative aimed at K-12 schools, we explore the intersection of chemical engineering and art through the captivating technique of water marbling. This project seeks to demystify the complexities of chemical engineering for young learners by relating fundamental concepts to a familiar and engaging artistic activity. Through the process of creating colorful patterns on the surface of water, which are then transferred to paper or fabric, students are introduced to a range of chemical engineering principles, including enzyme reactions, molecular mixing, surface tension, and fluid mechanics.

The core of this initiative lies in its hands-on approach, where students actively participate in water marbling while simultaneously learning about the underlying scientific principles. By demonstrating how specific changes in the water's composition or in the choice of materials lead to variations in the marbling effect, we illustrate the practical applications of chemical engineering in everyday life. This method not only makes abstract concepts more accessible but also sparks curiosity and enthusiasm for science and engineering disciplines.

This abstract outlines the development and execution of the outreach program, detailing the instructional strategies used to integrate art and science. Furthermore, it discusses the pedagogical framework for introducing complex engineering topics in an age-appropriate manner, emphasizing the role of creative activities in enhancing STEM education. Through this unique blending of art and engineering, we aim to broaden students' perceptions of what engineering encompasses and inspire a diverse generation of future engineers by highlighting the creativity and innovation at the heart of the field.

198g Flowing Futures: Empowering High-Schoolers in Chemical Engineering

Neha Raikar, Fernando Mérida Figueróa

Undergraduate Chemical Engineering (ChE) programs are currently experiencing declined enrollment due to factors including the dominant industry, alternative job offers, long graduation times, unstandardized curricula, and variable salary rewards. Misconceptions surrounding the ChE major as "abstract" or "polluting" are also contributors. Exposing students to experiential learning via hands-on activities earlier in their educational journey has the potential to boost enrollment and retention in the major. Outreach efforts are underway to encourage pre-college students to pursue chemical engineering.

A desk-scale fluid flow module has been developed and utilized by the University of Florida (UF) as an alternative to diversify the scales of experimentation in ChE laboratories and its potential use in remote learning. The University of Maryland, Baltimore County (UMBC) offers various week-long summer courses for high-school students under the Summer Enrichment Academy (SEA) umbrella. The "Introduction to Chemical Engineering" at the SEA aims to provide students with a preview of the exciting world of ChE while raising awareness of various career pathway opportunities.

In this work, we feature the collaboration of the two institutions that have partnered to seamlessly integrate the mini fluid flow module within SEA, aiming to expose high-school students to fluid flow phenomena, a cornerstone topic in ChE. The hands-on module, a testament to its effectiveness, consists of a 3D-printed fluidic bench featuring pipes of varying diameters and orientations, aquarium pumps, flexible tubing for flow connections, and a differential pressure sensor connected to an Arduino microprocessor. Students conduct differential pressure measurements across different flow paths, engaging in discussions with teammates and instructors to analyze the impact of pipe characteristics on pressure drop. The module's effectiveness is underscored by drawing parallels between small-scale fluid flow experiments and real-world pipe networks, ranging from industrial-scale manufacturing processes to microfluidic devices for biomedical applications.

The initial feedback from the first two iterations of the summer program is positive, indicating improved understanding and interest in chemical engineering among participants. Plans are currently underway to implement pre- and post-module surveys to qualitatively and quantitatively assess students' comprehension and interest in ChE, aiming to gauge any increase in self-reported familiarity with the various fluid flow topics such as flow rate, friction, pressure drop, viscosity, heat exchange, and other.

The collaborative efforts introduced in this work not only showcase broader strategies for enhancing STEM outreach but also emphasize the innovative nature of our partnership. This partnership enables us to pioneer new approaches via experimental demonstration, proven effective in engaging K-12 students and cultivating a dynamic learning environment for high-school students. This exciting prospect underscores our commitment to fostering innovation and long-term societal impact.

198h Integrating Service Learning at the University Level with a Summer Outreach Program to Increase High School Girls' Self-Efficacy in STEM

Geeta Verma, Sogol Asaei, Rebecca Ahn, Jacob Hostert, Julie N. Renner

Women earn around half of the total bachelor's degrees awarded in the U.S. but their share of degrees in the science and engineering disciplines is around 24%. Factors such as lack of encouragement, discrimination, lower self-efficacy, and bias toward their ability are some of the reasons that discourage women and girls from pursuing careers in STEM disciplines. Several strategies have been suggested to increase girls' interest in STEM careers. One approach is a summer outreach program for high school girls in which they get an opportunity to learn about science and engineering by performing hands-on activities under the guidance of college students who act as role models and mentors.

In the present work, a synergistic approach was implemented by combining service learning at the university level via an engineering project plan (EPP) module with a high school summer outreach program. The aim of this project was to increase the self-efficacy (belief about one's abilities) of students ranging from high school to graduate level, and positively impact the social responsibility attitudes of university student participants.

Validated survey results indicated an improvement in high school girls' confidence in learning and doing science after the summer outreach program. The EPP module helped college students improve their engineering design and project management skills. The college students also realized the importance of providing service to the community at the end of the module. Another benefit of the program is that the undergraduate students generated projects that were later used in the outreach program. Combining high-school community outreach with service learning had a positive impact on the self-efficacy of students from the high school to undergraduate and graduate levels, and had a positive impact on the social responsibility attitudes of university students.

460a State of the Institute

Darlene Schuster

Updates for department chairs on activities and initiatives of AIChE.

460b ABET Update

Randy Lewis

Information for department chairs regarding ABET accreditation will be provided.

460c Salary Survey

Tyler Johannes

The results of the latest chemical engineering faculty survey will be presented.

315a Being the Coach and Mentor While Training up the Next Generation of Chemical Engineers

Reginald Rogers Jr.

Presentation by the recipient of the 2023 Award for Excellence in Chemical Engineering Teaching Practice.

For innovative, caring, and supportive approaches to inclusive excellence in teaching and mentoring practices for all students.

315b Python Computing in the Chemical Engineering Curriculum

John Kitchin

Presentation by the recipient of the 2023 Award for Innovation in Chemical Engineering Education.

For contributions in computing education in chemical engineering through the creation of open-source Python libraries and course materials, social media and scientific publishing tools.

Computing in engineering education has continued to gain importance but it also remains a challenge to teach. The courses in computing taught in Computer Science do not typically cover topics like numerical methods, and usually lack domain specific applications. There is not a straightforward way to teach a class on numerical methods that is broadly applicable in engineering either. If it is too early in the curriculum students may lack the math and domain knowledge to understand them, and if it is too late students end up learning skills after they need them. I have spent the past 18 years learning to teach students how to use computing, specifically in using Python to solve science and engineering problems. In this talk I will talk about these challenges and how we have addressed them, and where remaining challenges in this educational space remain.

315c My Life As a Boomerang Chemical Engineer: Lessons and Implications from Engineering Role Identity Research

Allison Godwin

Presentation by the recipient of the 2023 Award for Research in Chemical Engineering Education.

For significant research contributions on engineering role identity development of diverse students and translation into evidence-based practices and policies for diversity, equity, and inclusion.

This talk will describe my research trajectory into engineering education research and back into a chemical and biomolecular engineering department and the influence these positions have had on the type of research I have pursued. I will discuss some of my early career fundamental research on engineering role identity and more recent efforts on effective pedagogies and practices to improve the climate of engineering and chemical engineering for women, Black, Latino/a/x, and Indigenous students, and individuals at the intersections of these and other marginalized identities.

315d Service Is Central: Lessons from Summer School and My Un-Sabbatical Year As Provost

Margot Vigeant

Presentation by the recipient of the 2023 Award for Service to Chemical Engineering Education.

For a lifetime career of dedicated service to chemical engineering education from session chair to ASEE/AIChE Summer School Chair.

461a Jeffrey a. Reimer – Scholar, Teacher, and Leader

Alexis T. Bell

In this presentation, I will summarize Jeff Reimer's accomplishments and contributions in three areas – as a scholar in the development and application of NMR techniques for the characterization of materials, as a teacher of chemical engineering principles, and as a leader through his service as department Chair. Having known Jeff from the start of his career at Berkeley, I have had the unique experience of working with him on research and holding many discussions with him regarding the stewardship of our department, the latter during some of the most challenging years of its life. I will also share my views on Jeff's interactions with the members of his research group, the students in his courses, and the members of our faculty community.

461b To Record or Not to Record – That Is the Question

Stuart Adler

Over the last 10 years, growing numbers of faculty in Chemical Engineering at the University of Washington have incorporated use of ***video media*** to enhance teaching and improve student access. This includes ***pre-recorded videos*** that students view as part of flipped classroom instruction. But more recently (initially in response to the COVID-19 pandemic), many of our faculty have adopted ***classroom recordings*** as a means to expand student access and provide a supplementary educational resource. However, due to concerns over dwindling attendance, as well as the effort required to make videos, many on our faculty have expressed *mixed opinions* about adopting these recordings as a general practice.

This talk describes our department's collective experience with the use of video media, including the motivations for using it, how we have evaluated its efficacy, processes we have used as a faculty to reach consensus on its general use, and what we have done to support faculty in creating classroom recordings that honor variations in teaching pedagogy. These experiences have led to a broad consensus among our faculty that for many of the courses we offer, video media provides students with a highly valuable resource that has overall positive educational benefits. These include expanded mechanisms for engagement, increased opportunities for active learning, higher levels of formative assessment and descriptive feedback, reduced student anxiety, and greatly enhanced equity, inclusion, and student access.

461c Flipping the Chemical Engineering Curriculum: Start with Design

Joseph Chada

The instructor of the first course in any curriculum plays an important role; they provide the 'first-look' to prospective students entering a new discipline. For the uninitiated, a first engineering course must balance teaching threshold concepts that often involve math, science, and problem solving to students with vastly different academic backgrounds. In recent years, UC Santa Barbara has overhauled its first-year chemical engineering course, drawing inspiration from Prof. Jeffrey Reimer's introductory textbook, to emphasize the creative design of new processes. We emphasize the importance of concepts, creativity in design, and authentic problem-solving rather than strictly numerical solutions. In this session, I will highlight our approach and its effect on improving both course outcomes and the student experience.

461d Jeff Reimer: A Lifelong Educator and Visionary in Magnetic Resonance Research

Raphaele Clement

Jeff Reimer's enthusiasm for pushing the boundaries of magnetic resonance technique continues to inspire many of us. This talk will provide an overview of Jeff's major accomplishments in the areas of spin hyperpolarization, solid-state NMR, and magnetic resonance imaging, with applications in spintronics, catalysis, and gas storage. Jeff's kindness and inclusivity will also be celebrated through anecdotes from students, colleagues, and friends who have had the good fortune of crossing his path.

461e Chemical Vapor Deposition of Conjugated Polymers for Flexible Lightweight Photovoltaics

Karen K. Gleason

Conjugated polymers exhibit promising mechanical properties crucial for the advancement of flexible energy and electronic devices. However, their widespread application necessitates precise molecular engineering to enhance electrical conductivity and carrier mobility. Indeed, the nanostructure and properties of conjugated polymers can be precisely controlled by chemical vapor deposition (CVD). The CVD conditions for semicrystalline poly(3,4-ethylene dioxythiophene) (PEDOT) have been tailored to achieve a face-on crystallite orientation rather than the more common edge-on orientation. The face-on orientation reduces the barrier to charge carrier transport between crystallites, thereby dramatically improving in-plane electrical conductivity. Additionally, the CVD method provides precise and systematic control over the b-axis lattice parameter (π-π stacking distance) within the crystallites. Decreasing the b-axis lattice parameter increases the charge transfer integral and enhances charge carrier mobility within the crystallites. Reducing the π–π stacking distance from 3.50 Å to 3.43 Å, yielding a dramatic electrical conductivity enhancement of $\approx$1140%. Improved electrical conductivity also correlates with superior morphological order, as indicated by reduced Urbach energies. A maximum electrical conductivity of 7520 ± 240 S cm^{-1} was achieved for CVD PEDOT. This conductivity exceeds that of the brittle inorganic transparent conductor indium tin oxide (ITO). Furthermore, the single-step, all-dry CVD growth process offers seamless integration onto various substrates, including thermally sensitive plastics, eliminating the need for complex transfer steps. This compatibility with flexible and wearable device substrates positions CVD PEDOT as a key enabler for diverse applications. CVD PEDOT has been successfully integrated into a variety of flexible devices, including photovoltaics. Such flexible and lightweight solar cells are desired for applications such as self-powered aviation, wearable electronics, and the Internet of Things (IoT). Integration of CVD PEDOT as a hole transport layer in inverted perovskite solar cells not only improved power conversion efficiency but also enhanced shelf-life stability. CVD PEDOT was combined with CVD-grown graphene, parylene, and PbS quantum dots to achieve flexible photovoltaics with a power conversion efficiency of 7.1% and a power-per-weight of 12.3 W g^{-1}. These findings underscore the immense potential of CVD PEDOT in shaping the future of flexible and lightweight energy solutions.

461f Liquid Crystalline and Crystalline Assembly of Discrete Oligomers

Emily Davidson

Here, we synthesize a series of precise, sequence-defined oligomers with backbones composed of main-chain mesogens bridged by flexible linkers. We find that these oligomers crystallize into sheets which then bend or twist to assemble into spherulites composed of scrolled lamellae, cylinders, or banded spherulites with locally defined chirality. These oligomers allow us to build a picture of crystallization that spans from a small-molecule description through structures typical of bent-core liquid crystal assembly to nearly polymeric crystallization into banded spherulites.

461g Engineering Peptide and Protein Nanostructures with Guidance from Solid-State NMR

Anant Paravastu

We use solid-state NMR methodologies developed for structural biology to investigate assemblies of designer peptide and peptoids. These assemblies, which mostly resemble amyloid-like cross-β structures or α-helical coiled coils, have been used for antigen presentation, extracellular matrix mimicry, and drug delivery. Rational design has made it possible to incorporate functional domains on nanofiber surfaces, modulate assembly pathways, and control assembly through molecular recognition. For known peptide assemblies, we have discovered several unexpected behaviors, including unexpected molecular structures, effects of bio-active domains on assembly pathways, a helix-to-sheet transformation, trans-to-cis backbone amide bond isomerization. Integrating computational design with what we have learned from investigating structures of known assemblies, we have recently performed increasingly complex rational design of better-controlled nanostructures.

461h Decoding Dynamics: Jeff Reimer's NMR Breakthroughs in Porous Material Metrology

Velencia Witherspoon

Jeffrey Reimer, a distinguished professor in the Department of Chemical and Biomolecular Engineering at the University of California, Berkeley, has significantly advanced the application of nuclear magnetic resonance (NMR) spectroscopy in studying transport and motion in porous materials. I will focus his and sometimes our pioneering work that leverages the intrinsic capabilities of MR to probe molecular dynamics, offering unprecedented insights into the behavior of molecules within complex, confined environments. By utilizing MR, Reimer has been able to elucidate diffusion mechanisms, adsorption processes, and the intricate interactions between fluids and porous media at a microscopic level. His research has profound implications for various fields where understanding the fundamental transport properties within porous structures is crucial. Reimer's innovative approach has not only enhanced the theoretical framework of NMR applications in material science but also paved the way for practical advancements in the design and optimization of porous materials for industrial applications. His contributions underscore the transformative impact of NMR spectroscopy in unveiling the dynamic processes within porous systems, thereby driving forward the frontiers of chemical engineering research.

461i Our Children, Our Students

Jeffrey A. Reimer

The poet Louise Gluck once wrote, "We look at the world once, in childhood. The rest is memory." It is certainly true that every generation experiences trauma and anxiety, memories that shape them. Yet the 21st century has been a wild ride for the millennials: from seeing their parents transfixed by crashing aircraft; their childhood marked with "global wars on terrorism;" at their fingertips they watch- if not experience- enormous calamities from a climate changing planet. Their teen years were held hostage by a world-wide epidemic; their media feeds punctuated with racial violence; their classrooms drilled regularly for gun violence.

They are now our students, both undergraduate and graduate, and they are acutely aware of social injustices documented on social media. Our students and their families are burdened with decades of debt for classroom content that often seems outdated, even when delivered via Zoom. It is no surprise, then, that there is a widespread crisis of student anxiety and depression, a concern voiced by many in the healthcare field. This is where our students and perhaps even our family members are at.

How do we do make chemical engineering education impactful in light of these realities? I propose that the answer lies in transforming and rebuilding our educational approach. This rebuilding means adapting to our current time and space and responding to students' physical, emotional, and psychological needs. I will illustrate this challenge of rebuilding with examples from my work as a chair, advisor, and teacher. These examples are meant to inspire, not prescribe, and I hope they will encourage you to use your considerable acumen and insight to develop strategies for rebuilding chemical engineering education for the 21st century.

122a Division Director Update on the Chemical, Bioengineering, Environmental, and Transport Systems Division of NSF: Programs and Opportunities

Jeanne VanBriesen

The mission of the Chemical, Bioengineering, Environmental and Transport Systems (CBET) Division of NSF's Directorate for Engineering is to support innovative research and education in the fields of chemical engineering, biotechnology, bioengineering, and environmental engineering, and in areas that involve the transformation and/or transport of matter and energy by chemical, thermal, or mechanical means. CBET research and education investments contribute significantly to the knowledge base and to the development of a 21st century workforce for major components of the U.S. economy, including chemicals, pharmaceuticals, medical devices, forest products, metals, petroleum, natural gas, food, textiles, energy utilities, alternative energy sources, microelectronics, and other sectors. Support for environmental research encompasses pollution prevention and remediation as well as life-cycle analysis.

The Division has four clusters of core programs for unsolicited proposals. The Chemical Process Systems (CPS) cluster of programs in CBET supports research seeking innovations in catalytic design, reaction engineering, multiphase chemical separations processes, molecular thermodynamics, process control and design, and sustainable energy conversion. The four programs are Catalysis, Interfacial Engineering, Electrochemical Systems, and Process Systems, Reaction Engineering, and Molecular Thermodynamics. The Transport Phenomena Cluster at CBET supports fundamental research on the transport of mass, momentum, energy, and species. There are four core programs in this cluster: Fluid Dynamics, Particulate and Multiphase Processes, Combustion and Fire systems, and Thermal Transport Processes. Research proposals supported by these core programs span across several disciplines including Chemical, Mechanical, Aerospace, Petroleum, and Environmental Engineering. The Engineering Biology and Health (EBH) cluster of programs in CBET supports a diverse research portfolio, ranging from nano-scale molecular design to macroscale assistive device development. Core programs in the EBH cluster include Cellular and Biochemical Engineering, Biosensing, Biophotonics, Engineering of Biomedical Systems, and Disability and Rehabilitation Engineering. The Environmental Engineering and Sustainability (EES) Cluster is organized around three core programs: 1) Environmental Engineering, 2) Environmental Sustainability, and 3) Nanoscale Interactions.

122b Proposal Writing Strategies for Fundamental Research Proposals for NSF

Dimitrios Papavassiliou

This workshop, conducted by a former program officer of NSF's Engineering Directorate, will discuss strategies for putting together fundamental research proposals, typical of the National Science Foundation. The workshop will discuss the importance of motivation of the problem, research approach, and describing the intellectual merit and broader impacts of the project. The workshop will be followed with informal networking with NSF program officers at the meeting.

58a The Right Reasons to Choose Industry over Academia and the Return

Gabriela Romero Uribe

Dr. Gabriela Romero is the Klesse Associate Professor in the Department of Biomedical Engineering and Chemical Engineering at the University of Texas at San Antonio. Dr. Romero received her B.S. in Chemical Engineering from the Universidad Autonoma de San Luis Potosi, Mexico in 2007, a M.S. in Advanced Materials Engineering in 2009 and Ph.D. in Applied Chemistry and Polymer Science in 2012 from the Universidad del Pais Vasco, Spain. During her graduate studies, Dr. Romero received twice the Marie Curie Fellowship to carried out research internships at Capsulution Pharma (Berlin, Germany) and at Zhejiang University (China). After completing her graduate studies, she held postdoctoral positions at the University of Kentucky (2012-2013) and at the Massachusetts Institute of Technology (2013-2015). Prior joining UTSA in 2017, she worked for two years as a Senior Scientist at Poseida Therapeutics (La Jolla, CA). Dr. Romero's primary research interests are the investigation of biomedical soft materials and hybrid platforms as therapeutic tools for the treatment of brain diseases. She has published numerous scientific papers in prestigious journals, including Science, Nature, Advance Functional Materials, ACS Advanced Materials & Interfaces, and Soft Matter. Dr. Romero has nearly 3 million dollars in research funding from federal agencies such as the National Institute of Neurological Disorders and Stroke, the National Institute of General Medical Sciences, National Institute of Biomedical Imaging and Bioengineering, and the National Science Foundation, including the NSF CAREER award. Dr. Romero's academic career philosophy is capitalized in supporting her continued development as an advocate for minority participation in advanced science and engineering education, particularly the participation and mentoring of Latinas in Engineering.

58b A to B Via C: Professional Detours before Arriving in Academia

Keisha Antoine

Dr. Keisha Antoine began her professional journey at Lehigh University, where she obtained a B.S. in Chemical Engineering in 2001 and a M. Eng. and PhD in Materials Science & Engineering in 2004 and 2007, respectively. She is a registered professional engineer in the state of TX and is currently a Lecturer I in the Chemical Engineering Department at Prairie View A&M University. Her research interests are interdisciplinary in nature involving the use of immersive technologies in engineering education and the study of tropical green foods with a focus on food security. Prior to her involvement in academia, Dr. Antoine had a successful career as a scientist/engineer at Corning Incorporated, where she was granted 3 patents for her work on glass processes and functionalization.

58c Career Transitions of Professor Kintu Early

Kintu Early

Kintu is a chemical engineer with an extensive background in materials and materials processing. Early received his B.S. in Chemical Engineering from Northwestern University and his Ph.D. in Chemical Engineering from University of Pittsburgh. Following graduate school, he began working as a development engineer with Corning Incorporated in Optical Communications where he worked for twenty three years before coming to Cornell in August of 2023 as a Professor of Practice and Director of the Master of Engineering Program in Materials Science.

58d Late-Stage Industry to Academia Career Transition

Jean Tom

After a 38-year career in the pharmaceutical industry, Jean Tom is embarking on an academic career as a teaching faculty. She was the 2024 Brenton Halsey Distinguished Visiting Professor of Chemical Engineering at the University of Virginia and is currently a Professor of Practice at Princeton University's Department of Chemical and Biological Engineering. Jean recently retired from Bristol Myers Squibb, where she served as Executive Director of Development Engineering - Chemistry Process Development. In this role, she led a group of chemical engineers focused on developing chemical processes to synthesize small-molecule drug candidates for new therapeutics. Before joining BMS in 2006, Jean spent 19 years at Merck Research Laboratories.

In her industrial role, Jean has been a strong advocate for increasing the impact of chemical engineering in the pharmaceutical industry. She introduced new approaches to accelerate the generation of process knowledge needed at different stages of development and promoted pre-competitive collaboration to advance new technologies for developing new compounds. Jean is actively involved in her profession through the AIChE (American Institute of Chemical Engineers), CACHE (Computer Aids for Chemical Engineering), and the NAE (National Academy of Engineering). She also contributes to the chemical engineering educational community as an ABET program evaluator, an external advisory committee member for several departments, and through STEM outreach programs.

199a New Funding Opportunities at NSF of Interest to the Aiche Research Community

Carole Read, Christina Payne, Steven Peretti, Shahab Shojaei-Zadeh, Carl Rockne

This workshop will highlight Engineering Directorate and cross Directorate funding opportunities at the National Science Foundation that may be of interest to the AIChE Research communities. We will also discuss programs of NSF's newest directorate, Technology, Innovation and Partnerships (TIP). Example programs including Future Manufacturing, Accelerating Innovations in Biomanufacturing with the DOE's Agile BioFoundry, the NSF/CASIS Collaboration on Transport Phenomena Research on the International Space Station.

199b Epscor Program Funding Opportunities

Hongmei Luo

The Established Program to Stimulate Competitive Research (EPSCoR) is a part of the U.S. NSF's portfolio that pursues a mission to enhance the research competitiveness of targeted jurisdictions by strengthening science, technology, engineering, and mathematics (STEM) capacity and capability through a diverse portfolio of investments. This talk will discuss EPSCoR program funding opportunities for the entire EPSCoR community.

199c Panel Discussion with NSF Program Officers

Carole Read, Christina Payne, Steven Peretti, Catherine Walker, Shahab Shojaei-Zadeh, Carl Rockne

This talk will be an open discussion with NSF CBET Program Officers on Core Programs and cross Engineering Directorate and Cross Foundation funding opportunities.

Combining Ontologies and Llms to Evaluate and Improve Engineering Curricula

Nikolaos Trokanas, Madeleine Bussemaker, Antonis, Kokossis, Franjo Cecelja
e-mail:n_trokanas@mail.ntua.gr

This research introduces an innovative framework to evaluate and enhance engineering curricula by integrating domain-specific ontologies with Large Language Models (LLMs). As engineering fields continuously evolve, curricula must keep pace with the rapid advancement of technologies and industry needs. Traditional curriculum evaluation methods rely on static criteria and limited interdisciplinary perspectives, often failing to capture the complexities of modern engineering requirements. This study addresses these challenges by creating a more adaptive, comprehensive, and data-driven approach to evaluating and updating engineering programs.

The methodology begins with data collection and preprocessing from multiple sources to ensure consistency in curriculum evaluation across different programs and institutions. By standardizing and preprocessing data, the research prepares a solid foundation for building domain-specific ontologies that reflect the core knowledge in various engineering fields. These ontologies are integral to the project, providing a structured representation of essential concepts, from fundamental engineering principles to emerging topics such as nanotechnology, bioprocessing, and interdisciplinary fields like data science and sustainability. Once established, these ontologies are integrated with state-of-the-art LLMs, enabling the system to bridge structured and unstructured data sources. LLMs are fine-tuned to identify gaps, clarify ambiguous concepts, and suggest improvements in curriculum content. The integration allows for dynamic knowledge representation, in which the ontologies provide a stable framework while the LLMs adapt to evolving industry standards and interdisciplinary trends. This approach facilitates the identification of gaps and redundancies in course content and pinpoints areas that may benefit from updated or additional content, such as recent advancements in technology or regulatory changes. The curriculum mapping and analysis phase leverages these ontologies and LLMs to create a knowledge graph that represents the curriculum in terms of key concepts and their relationships. This mapping enables the system to evaluate the curriculum's content coverage by identifying missing concepts, overlaps, and potential interdisciplinary connections that could enrich students' learning experiences. Specifically, the LLMs analyze course materials against current industry requirements and identify misalignments between the curriculum and the latest technological and regulatory demands. A focused use case within the Chemical Engineering curriculum demonstrates the effectiveness of this methodology. By applying a comprehensive ontology, CheEdO, that includes core concepts such as learning objectives, modules, credits, as well as scientific concepts such as mass transfer, thermodynamics, safety standards, and process design, the framework analyzes the curriculum's alignment with industry requirements. Through gap analysis, the system identifies areas lacking in content or needing updates and recommends additions to maintain relevance with industry expectations. For instance, the system may suggest incorporating interdisciplinary topics like sustainability or data science to better prepare students for modern engineering challenges. This combined ontology-LLM approach also offers data-driven insights and recommendations to curriculum designers. The fine-tuned LLMs analyze broader scientific and engineering literature to propose potential interdisciplinary connections and emerging topics, such as courses on sustainable engineering practices or advanced computational methods. As a result, the system generates actionable suggestions for curriculum improvements, including new courses, expanded content in existing courses, and updates to outdated materials.

Preliminary results indicate that this hybrid approach is highly effective in evaluating engineering curricula by uncovering knowledge gaps and redundancies while generating relevant insights for integrating new technologies and interdisciplinary knowledge. By enhancing curricula with dynamic, adaptable, and data-driven recommendations, this methodology ensures that engineering programs remain effective and relevant in preparing students for the demands of modern engineering careers. This research thus provides a significant advancement in curriculum evaluation, aligning educational programs with evolving industry standards and positioning students to excel in an increasingly complex engineering landscape.

This project has received funding from the European Union's Horizon 2020 research and innovation programme BIOINDUSTRY 4.0 (grant agreement 101094287)

376b Radical Redesign of an Introductory Chemical and Biomolecular Engineering Course for Student Motivation

Jostin Armada, Spencer Sabatino, Trevor Franklin, Allison Godwin

Many introductory chemical engineering courses use traditional approaches to teaching core principles in the field (e.g., material balances, process flow diagrams, energy balances, etc.) through abstract problem-solving scenarios, divorced from real-world contexts, and focused on typical industrial applications. As a result, chemical engineering is often introduced in ways that are unconnected to students' motivations for studying engineering and their lived experiences. These trends were true of an introductory chemical and biomolecular engineering course taken by first-semester, first-year students at a private, Northeastern research-intensive institution. Additionally, student enrollment in the course had declined in the previous four years (consistent with national trends of declining chemical engineering enrollments; Petruzzeli, 2022), and approximately half of the students who completed the course ultimately chose majors outside of chemical engineering.

The course was redesigned in Fall 2023 through an institutional grant to focus on 1) hands-on, active problem-solving, 2) mastery-based grading, and 3) teaching core chemical engineering concepts through food as ways to engage students' personal motivations to learn chemical engineering and pursue chemical engineering degree pathways. This three-prong approach leveraged the significant body of literature on the impact of active learning techniques on student learning (Freeman et al., 2014) and motivation (Stolk et al., 2018; Zengaro & Zengaro, 2022), the emerging scholarship on the possibilities of ungrading methods for lowering student stress and increasing engagement with course materials (Landherr, 2023), and the use of food as a more compelling and relatable educational context for teaching chemical engineering than traditional methods (Carvalho do Prado, 2024; Vigeant, 2021).

The redesign leveraged Problem-Based Learning (PBL) and Goal Orientation Theory as pedagogical and research frameworks, respectively. PBL enables students to actively construct novel knowledge rooted in both current course content and prior experiences (Fierra & Trudel, 2012). PBL fosters independent learning by prompting students to seek need-to-know information pertinent to the problem context (Ertmer & Glazewski, 2019). Goal Orientation was used to understand student motivation for engagement in course content and pursuit of chemical engineering as a major. Goal orientations refer to the reasons or purposes for engaging in learning activities and explain individuals' different ways of approaching and responding to achievement situations (Ames & Archer, 1988). We focus on two orientations linked to student learning outcomes: 1) performance approach (i.e., goals focused on the demonstration of competence relative to others or prior self-performance) and 2) mastery approach (i.e., goals related to improving knowledge, skills, and learning; Elliot & McGregor, 2001).

The traditional course consisted of two 50-minute lecture periods per week with one 110-minute discussion section for teaching assistant-guided problem-solving. The grading scheme included two exams each worth 25% of the final grade; a team project worth 35% of the final grade; and class participation, discussion section attendance, and other designated assignments that accounted for 15% of the final grade. All assessments were summative. Assessment methodologies wield considerable influence over students' motivational states, either positively or negatively. Traditional grading systems often perpetuate notions of fixed intelligence, detrimentally impacting student motivation (Malespina, 2022). Alternatively, adopting grading frameworks centered on mastery, feedback provision, and iterative improvement can significantly bolster student motivation, cultivate a growth mindset regarding learning, enhance overall well-being, and improve academic outcomes (Hsu & Goldsmith, 2021).

In the course redesign, the contact time and meeting structure remained the same. However, the discussion sections were transformed into hands-on labs, connected to the concepts being learned in the course. All of the course content was taught through food examples and included material balances with and without reaction, simple vapor-liquid equilibrium, simple energy balances, biomolecular engineering, the social impact of food, scale-up, safety, separations, and engineering design. Grading was changed to a mastery-based grading scheme. All problems were graded as well-developed (i.e., virtually no errors), developing (i.e., minor conceptual errors that affected the quality of the solution), under-developed (i.e., major conceptual errors that affected the quality of the solution), and no evidence (i.e., no work completed or insufficient work to gauge student understanding of the learning objectives). Students were assigned weekly homework or design milestones (graded formatively), one midterm exam (graded summatively), pre-labs (graded for completion before labs), a final design project presentation and report (graded summatively), and additional reflection assignments. For formatively graded assignments, students could revise the submission within one week of feedback and a reflection on the source of the error(s) and plan to address them in future assignments. Students who earned under-developed or developing scores could receive up to a well-developed score, and students who earned no evidence score could receive up to a developing score.

Students proposed their final grades in the course. Grading guidelines were negotiated on the first day of class to set expectations for "grade bins"; for example, an A grade bin in the course required > 85% well-developed on the exam, > 90% well-developed on the final project deliverables, and > 90% well-developed on the homework/design milestones OR > 75% well-developed on the exam, > 90% well-developed on the final project deliverables, and > 95% well-developed on the homework/design milestones (with 0 no evidence scores) AND 100% well-developed on the pre-labs and completion of all teaming and reflection assignments. Students completed a guided reflection on their mastery in each category, which grade bin their performance matched with, and a proposed grade and rationale. If any disagreement occurred between the proposed score from the student and the instructor's expectations, a short Zoom meeting was scheduled to discuss.

The redesigned course was implemented in Fall 2023 with 48 students. The teaching team consisted of two co-instructors, a graduate teaching assistant, and three undergraduate teaching assistants. To understand the impacts of the curricular change on student motivation and career intentions, multiple sources of data were collected, including pre/post-surveys of student attitudes, career intentions, and understanding of chemical engineering, which were analyzed with a paired samples Wilcoxon test. Course assignments, including reflections, assignments, and engagement in the revision process, were also tracked. Course evaluations also provided input on the course outcomes. Finally, a graduate student outside of the teaching team conducted one-hour semi-structured interviews with five students about their experiences, which were coded inductively.

Results indicate that the course grade distribution was not statistically significant from prior semesters and that the new structure challenged students while reducing stress. The course supported a mastery goal orientation environment, and 89% of the class redid more than half of the formative assignments. Students reported a deep understanding of chemical engineering concepts and a broad understanding of possible future career pathways. Opportunities for improving this course delivery included a clearer grading schema and stronger training of teaching assistants for implementing the course vision. Overall, this approach improved student motivation and supported students' decision-making for chemical engineering major pathways.

References

Ames, C., & Archer, J. (1988). Achievement goals in the classroom: Students' learning strategies and motivation processes. *Journal of Educational Psychology*, *80*(3), 260–267. https://doi.org/10.1037/0022-0663.80.3.260

Carvalho do Prado, G. H. (2024). Teaching food processing to chemical engineering students with Native American food. *Chemical Engineering Education*, *58*(1), 46-53. https://doi.org/10.18260/2-1-370.660-132143

Freeman, S., Eddy, S. L., McDonough, M., Smith, M. K., Okoroafor, N., Jordt, H., & Wenderoth, M. P. (2014). Active learning increases student performance in science, engineering, and mathematics. Proceedings of the National Academy of Sciences, 111(23), 8410-8415. https://doi.org/10.1073/pnas.1319030111

Ferreira, M. M., & Trudel, A. R. (2012). The impact of problem-based learning (PBL) on student attitudes toward science, problem-solving skills, and sense of community in the classroom. Journal of Classroom Interaction, 23-30. https://www.jstor.org/stable/43858871

Elliot, A. J., & McGregor, H. A. (2001). A 2x2 achievement goal framework. *Journal of Personality and Social Psychology*, *80*(3), 501–519. https://doi.org/10.1037/0022-3514.80.3.501

Ertmer, P. A., & Glazewski, K. D. (2019). Scaffolding in PBL environments: Structuring and problematizing relevant task features. In M. Moallem, W. Hung, & N. Dabbagah (Eds.), The Wiley Handbook of Problem-Based Learning (pp. 321-342). Wiley.

Landherr, L. (2023). Ungrading in chemical engineering: Attempting to eliminate exams, deadlines, and anxiety by refocusing on learning instead of grades. In proceedings of the *American Society for Engineering Education Annual Conference & Exposition*, Baltimore, MD. https://doi.org/10.18260/1-2--44552

Malespina, A., Schunn, C. D., & Singh, C. (2022). Whose ability and growth matter? Gender, mindset and performance in physics. International Journal of STEM Education, 9(1), 28. https://doi.org/10.1186/s40594-022-00342-2

Hsu, J. L., & Goldsmith, G. R. (2021). Instructor strategies to alleviate stress and anxiety among college and university STEM students. CBE—Life Sciences Education, 20(1). https://doi.org/10.1187/cbe.20-08-0189

Petruzzeli, E. (2022, March). Editorial: Concerning trends for ChEs. *CEP Magazine*. https://www.aiche.org/resources/publications/cep/2022/march/editorial-concerning-trends-ches

Stolk, J. D., & Zastavker, Y. V., & Gross, M. D. (2018). Gender, motivation, and pedagogy in the STEM classroom: A quantitative characterization. In proceedings of the *American Society for Engineering Education Annual Conference & Exposition*, Salt Lake City, UT. https://doi.org/10.18260/1-2--30556

Vigeant, M. (2021). The oldest unit operation. *Chemical Engineering Education*, *55*(1), 9-10. https://doi.org/10.18260/2-1-370.660-125292

Zengaro, S., & Zengaro, F. (2022). Active learning, student engagement, and motivation: The importance of caring behaviors in teaching. In *Handbook of Research on Active Learning and Student*

Engagement in Higher Education (pp. 66-83). IGI Global.

376c Student Involvement in Generating Material for a Hands-on Brew Lab: Reflections on Self-Efficacy and Engagement

Joshua Abraham, Brian Pfleger

Studies have shown that student-generated content can enhance learning by centering curriculum around the learner: this creates space for authentic exploration of topics intriguing to the learner and can build rapport between the instructor and their students. We adopted this paradigm when building out the new Nanobrewery and Fermentation Instructional Laboratory at the University of Wisconsin, Madison. Four undergraduates worked on independent projects where they identified unit operations within the brewing process, brainstormed potential variables and measurable outcomes, and tested experiments to generate a draft protocol that would be usable in future courses. Following this experience, we implemented a similar, student-driven process for about 20 undergraduate students in a summer laboratory course to design and carry out experiments of their choosing. Students were asked to self-report their findings and reflect on their experience during this process. We anticipate that building flexibility into student engagement increases student self-efficacy, or their confidence to complete tasks relevant to the brew lab objectives.

376d Modeling Drug Delivery Using Sodium Alginate Beads and Food Coloring As an Intro to Chemical Engineering Experiment for First-Year Students

Alex Bertuccio

At the University of Kansas (KU), engineering students declare their major prior to the start of school. In their first year, students take an Intro to Engineering course in their respective discipline to learn more about the major and to decide if they want to further pursue their chosen major or switch. At KU, the Introduction to Chemical Engineering course is two one-credit courses that meet weekly for 50 minutes and students complete four two-hour lab experiments each semester. The topics cover a wide range of fields where chemical engineering is applied.

One experiment that was piloted in the second part of the sequence was a drug delivery experiment where students worked in groups and made sodium alginate beads that encapsulate food coloring. The food coloring is used a visual, representative drug. Prior to lab, students were given a procedure on how to make the beads and tasked with selecting and testing the effects of one variable on food coloring release for one hour. A spectrophotometer was used to quantify the amount of food coloring released. As students waited for the food coloring to be released, they developed their own calibration curve to relate concentration with absorbance.

Students gained experience designing an experiment, learned about the theory and calculations related to spectrophotometry, created their own calibration curve, and practiced performing dilutions and pipetting. This talk will focus on the format of the experiment, the learning objectives and showcase some results students obtained.

376e Enhancing Engineering Education: Low-Cost Hands-on Fluidized Bed Heterogeneous Reactor with Color Change

Faraz Rahimi, David B. Thiessen, Riley Fosbre, Prashanta Dutta, Olusola Adesope, Bernard Van Wie

Fluidization is a pivotal process in chemical engineering, involving the suspension of solid particles in a fluid medium to achieve enhanced mixing and reaction rates. In this study, we introduce a novel low-cost desktop learning module designed to simulate fluidization with a reaction, marking a significant advancement over previous modules that lacked reaction capabilities. Our objective is to develop a classroom-friendly tool operable under ambient conditions, ensuring safety by employing non-toxic materials. By utilizing an enzyme catalyst, we aim to induce a visible color change through a reaction with sugar, enhancing the educational experience by providing a tangible outcome. Integrating such hands-on experiments promotes active learning and allows students to observe real-world phenomena in controlled settings, fostering a deeper understanding of chemical engineering principles.

Furthermore, implementing low-cost desktop learning modules offers numerous advantages for student learning. First, it enhances accessibility, allowing students to engage with complex concepts outside traditional laboratory settings. Moreover, the student team interactive nature in which these modules are implemented promotes self-directed learning and problem-solving skills, empowering students to explore concepts at their own pace. Additionally, using non-toxic materials ensures safety in classroom environments, alleviating concerns associated with traditional experimental setups.

In the event of unforeseen challenges or limitations in our methodology, we remain committed to exploring alternative approaches to achieve our educational objectives. Through continuous refinement and innovation, we aim to create effective tools that inspire and educate future generations of chemical engineers.

376f A New Way of Teaching Gas Liquid Reaction Kinetics and Gas-Liquid Reactor Design

Damaraju Phaneswararao

The topic of gas-liquid reactions and reactor design is recommended to be included in the course contents of a second course on Chemical Reaction Engineering at a senior UG level either as an elective or compulsory course in some schools or in the course content of a post graduate level PG course. The topic of gas-liquid reactions is dealt to a varying degree of extent in several text books that are available in the literature. the prominent books among them being those authored by Astarita(1967),Astarita et al (1983), Danckwerts(1970), Doraiswamy and Sharma(1984), Westerterp et al (1991), Levenpsiel (1972,1984,1999), Froment and Bischoff(2011). The books of Astarita (1967), Astarita et al (1983) and Danckwerts (1970) delve into the theory of gas liquid reactions and summarize the relevant literature available till the date of the publication of the respective book but they address the researcher rather than a class room teacher or student. The text book of Doraiswamy and Sharma (1984) explains the theory and applications of gas liquid reactions quite well but does not enable the teacher to train the student in problem solving with respect to identifying rate of a gas-liquid reaction and writing and solving the design of a chosen type of gas-liquid reactor. The Chemical reactor Omnibook (1984) discusses solution of gas-liquid reactor design problems by using the enhancement factor vs Hatta Modulus plot of Van Krevelen and Hoftijzer (1954) for a (1,1) order reaction taking place in a liquid film but which is obtained for the case where the reaction in bulk is assumed to reach completion.. No examples are given to illustrate the use of the equations but a host of problems are assigned at the end of the chapter on gas liquid reactions some of which may require solution of the film model for the case where reaction in bulk may not reach completion or where there is a finite flux of the solute into the liquid bulk. Neither the second or the third edition of the book by Levenspiel contain good number of examples. The importance of the role of gas film resistance in finding the rate of absorption is emphasized quite well in the book by Levenspiel unlike other text books but the procedures for taking gas film resistance into account require trial and error. Same way the equations required for design of ideal reactors are given with no illustrations but plenty of problems are assigned to the student which require lot more analysis and new derivations which are not hinted in the book. The text book of Westerterp et al (1991) covers the theory of absorption with reaction quite well but unfortunately the theories for gas-liquid reactions and solid catalyzed reactions are dealt in the same chapter even alternately which creates lot of confusion to the reader and teacher when he has to present the text book material in the class room. The analogy between the diffusion reaction equations for the two types of reaction systems is restricted and hence it is easier and simpler if they are treated separately. The text book of Froment and Bischoff (2011) treats the kinetics and design of gas-liquid reactions in separate chapters with some illustrations but is not exhaustive enough to enable the theory given by them to solve problems from the Chemical Reactor Omnibook (1984) of Levenspiel. It is rather surprising that in spite of the fact that research on the theory of gas-liquid reactions started from the application of Whittman's film theory as early as 1928 (Hatta, 1928,1935) or so, research papers appeared in chemical engineering science journal up to 2018(Stepanek et al (2018)) or so discussing the equations for predicting enhancement factor for the simplest case of a (1,1) order irreversible gas liquid reaction. Thus, when it comes to teaching of the topic on kinetics and design of gas-liquid reactions and reactors, all of the above mentioned books appear to be lacking in one respect or another when it comes to imparting the student the ability to tackle problems relating to development of rate equations and reactor design procedures. But we find that relatively a lot more coverage is given to the theory of solid catalyzed gas phase or gas-liquid reactions and reactors or the theory of gas-solid(reactant) reactions and reactors.

In the present work, rate equation for the (1,1) order irreversible gas liquid reaction are presented which consider the most general case where the gas film resistance, reaction in film and bulk, depletion of the liquid phase reactant B in the film are derived and from this most general case by extending the Hikita and Asai (1964) approximation as done by Bhattacharya et al (1987).. Rate equations and criteria for the applicability of these special cases, where not all the resistances mentioned above are important, i.e cases corresponding to different regimes of absorption , are derived. The simple notation of these special cases introduced by Doraiswamy and Sharma (1984) is followed in the present work with a slightly modified definition of regimes and naming the gas film resistance controlled mass transfer as an additional regime. Accordingly , the following classification of regimes is followed in the present work. Regime 1 is defined as the regime of absorption with reaction where gas film resistance is negligible, reaction in film is insignificant, liquid phase mass transfer resistance is negligible and reaction takes place only in the reactor bulk and is not complete in the bulk. Regime 2 is defined as the regime where gas film resistance is negligible, reaction in film is insignificant, reaction in film is complete and the controlling resistance is only the liquid phase mass transfer. Regime 3 is defined as the regime in which gas film resistance is insignificant, reaction in film is complete and there is no flux of the solute from the film into the liquid bulk. Regime 4 is defined as the regime in which gas film resistance is not significant, reaction in film is purely diffusion controlled i.e reaction between the solute and the solvent is instantaneous. Regime 5 is defined as the regime in which gas film resistance is controlling and all other resistances are negligible compared to the gas film resistance irrespective of the other features regarding the zone of reaction i.e film or bulk. Mixed regimes where the features of one or more individual pure regimes mentioned above are to be considered as significant such as regime 1 and 2, regime 1,2, and 3, regime 3, 4, and 5, so on. The equations and criteria for the pure or mixed regimes are derived after developing at first , the rate equation for the most general case where features of all the pure regimes mentioned above are important. From the equations for the most general case, rate equations and the criteria for the validity of the various pure and mixed regimes are derived. The need to identify the controlling resistances arises from the fact that the selection of the reactor type depends on which hydrodynamic parameter is involved in the rate equation for the controlling regime that is identified based on input data for a given problem. The above approach has been used by the author while teaching Chemical Reaction Engineering at the senior undergraduate and post graduate level at Indian Institute of Technology (Delhi).over two decades. Also, application of the above approach to solve all the problems assigned in the Chemical Reactor Omnibook was developed and tested in the classroom and a broader set of exercises are also developed to enable the student to develop rate equations and criteria for new reaction systems.

REFERENCES

1. Astarita, G. Mass transfer with chemical reaction, Elsevier, Amsterdam, 1967

2. Astarita, G.; Savage,;D. W.; and Bisio, A. Gas treating with chemical solvent; John Wiley &amp Sons, New York, 1983.

3. Danckwerts, P.V., Gas liquid reactions. McGraw- Hill, London, 1970

4. Doraiswamy K.; Sharma, M.M., Heterogeneous Reactions Vol. 2, Wiley. New York, 1984.

5. Westerterp , K. R.; Van Swaaij, W. P. M.; Beenackers, A. A. C. M., Chemical Reactor Design and Operation. 2 nd edition, John Wiley &sons , NY, 1991

6. Levenspiel, O., Chemical Reaction Engineering, 2nd Wiley, NY. 1972

7. Levenspiel, O., Chemical Reactor Omnibook. Oregon book store, Corvallis, Oregon, 1987
8. Levenspiel , O., Chemical Reaction Engineering. 3rd ed. Wiley, NY, 1999.
9. Froment, G. F.; Bischoff, K.B.; De Wilde, J., Chemical Reactor Analysis and Design,Wile, 2011.
10. Van Krevelen, D. W.; Hoftijzer, P. J., Trans. I. Chem. E. 32, 5360, 1954
11. Hatta, S., Reprs. Tohoku Imp. Univ. 8, 1, 1928.
12. Hatta, S., Tech. Reprs. Tohoku Imp. Univ. 10, 119, 1932.
13. Stepanek, V.; Palaty, Z.; and Bendova , H., A critical comparison of approximate solutions for enhancement factor for irreversible second-order reaction. Chem. Eng. Sci.191, 410-419, 2018. https://doi.org/10.1016/j.ces.2018.07.003
14. Hikita, H.; and Asai, S. Gas Absorption with (m, n)-th Order Irreversible Chemical Reaction. Int. Chem. Eng. 4, 332, 1964
15. Bhattacharya, A.; Gholap, R. V,; Chaudhari, R. V. Gas absorption with exothermic bimolecular (1,1) order reaction. AIChE J. 1987,33(9),1507-1513.DOI.org:10.1002/aic.690330910

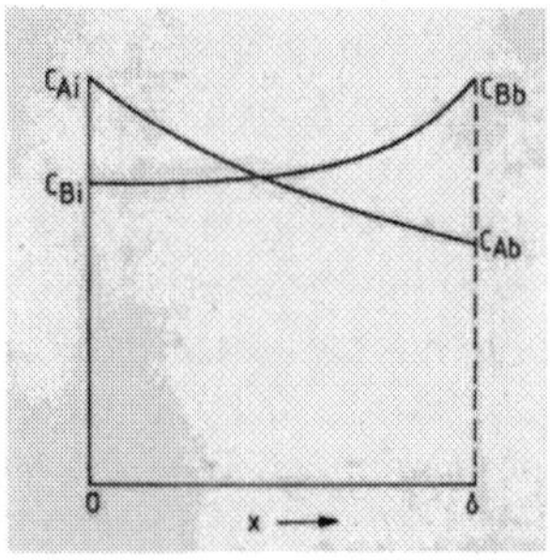

REGIME 1-2-3-4-5; REACTION IN FILM AND BULK($\frac{C_{Ai}H_A}{p_A} < 1$; $\frac{C_{Ab}H_A}{p_A}$>0

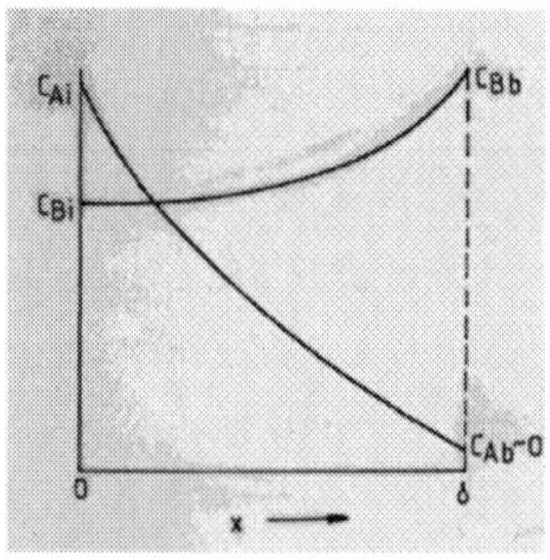

REGIME 2-3-4-5 REACTION IN FILM, REACTION COMPLETE IN BULK ($\frac{C_{Ai}H_A}{p_A} < 1$; $\frac{C_{Ab}H_A}{p_A}$>0)

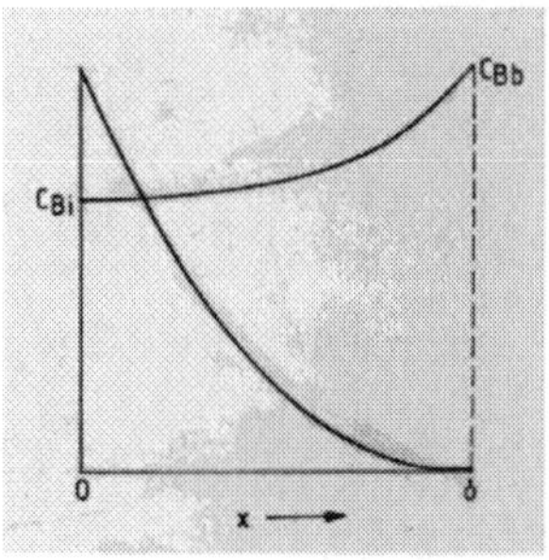

REGIME 3-4-5: REACTION IN FILM; NO FLUX OF SOLUTE INTO BULK;

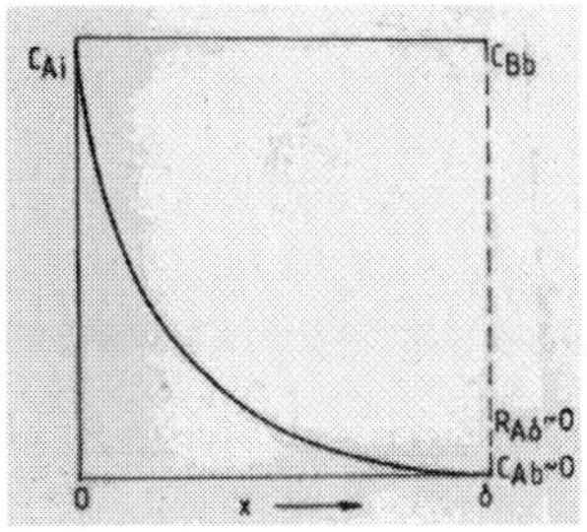

REGIME 3-5: REACTION IN FILM; NO DEPLETION OF B IN FILM; NO FLUX OF SOLUTE INTO BULK

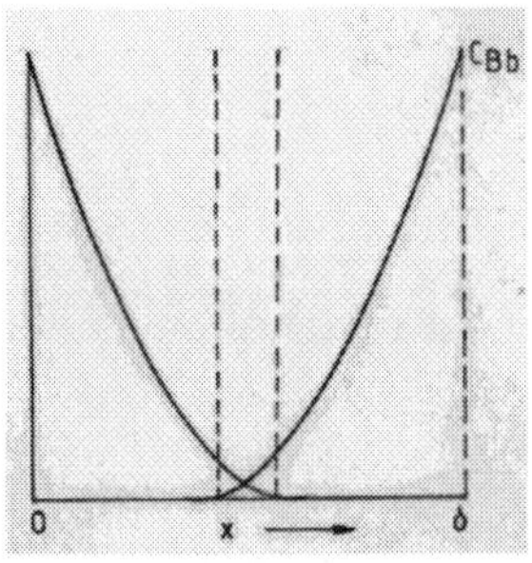

REGIME 4-5: DIFFUSION CONTROLLED REACTION IN FILM;

REACTION ZONE THICKNESS ≈0

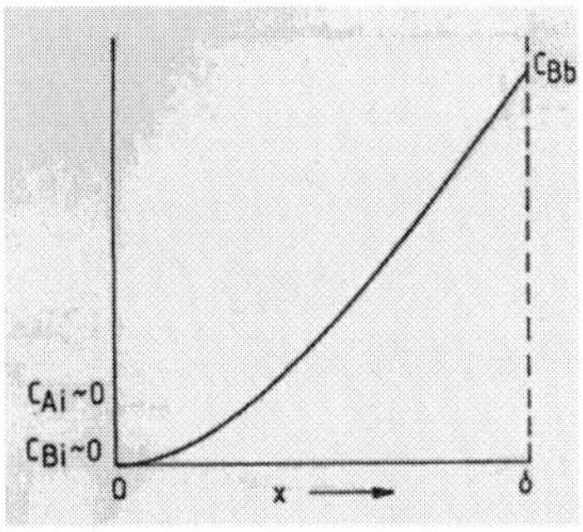

REGIME 5: GAS FILM CONTROLLED INSTANTANEOUS REACTION REGIME

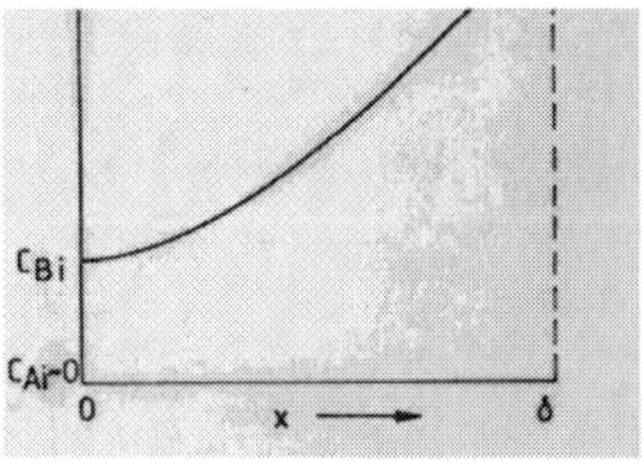

REGIME 5: GAS FILM CONTROLLED INSTANTANEOUS REACTION REGIME (p>1)

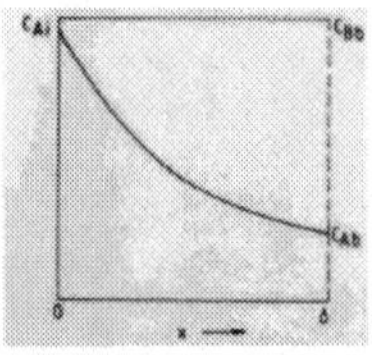

REGIME 1-2-3-5: REACTION IN FILM AND BULK ; NO DEPLETION OF B IN FILM

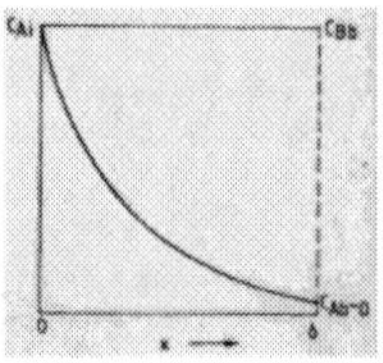

REGIME 2-3-5: REACTION IN FILM; REACTION COMPLETE IN BULK; NO DEPLETION OF B IN FILM;

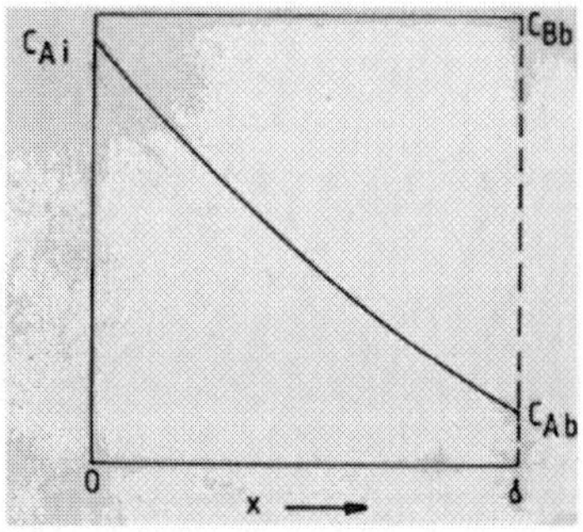

REGIME 1-2-5: NO SIGNIFICANT REACTION IN FILM; REACTION IN BULK AND DOES NOT REACH COMPLETION;

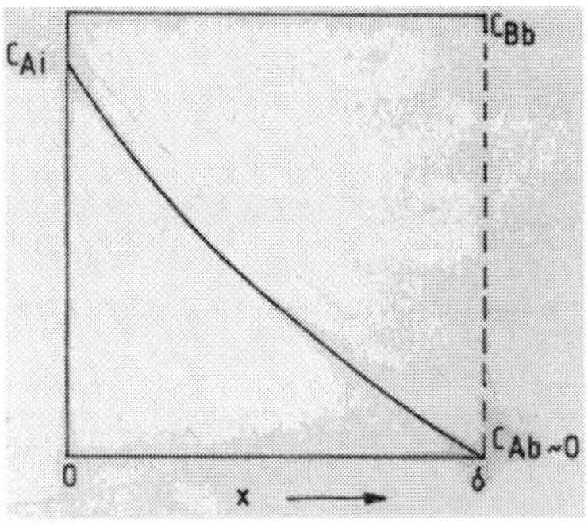

REGIME 2-5: NO SIGNIFICANT REACTION IN FILM; REACTION COMPLETE IN BULK;

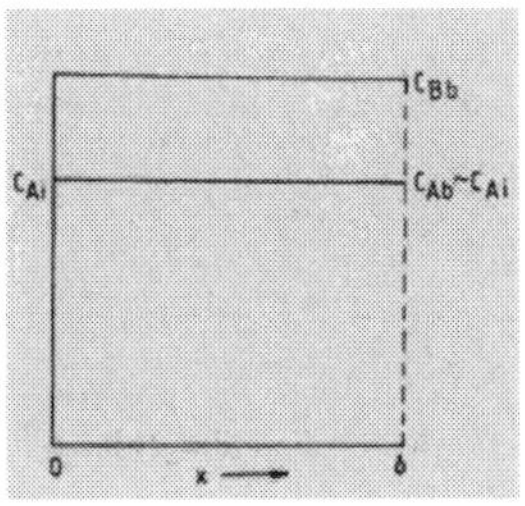

REGIME 1: REACTION ONLY IN BULK AND NO SIGNIFICANT GAS FILM OR LIQUID FILM RESISTANCE

$$\overline{C_{Ai}} = \frac{t_2}{1 + t_2 - \frac{1}{\cosh^2\sqrt{M}\eta \cdot \left[1 + \frac{R_2}{R_1\varphi_2}\right]}} \tag{1}$$

$$\overline{C_{Ab}} = \frac{\overline{C_{Ai}}}{\left[1 + \frac{R_2}{R_1\varphi_2}\right]\cosh\sqrt{M}\eta} \tag{2}$$

$$\overline{R_A} = \varphi_2\overline{C_{Ai}}\left[1 - \frac{1}{oosh^2\sqrt{M}\eta \cdot \left[1 + \frac{R_2}{R_1\varphi_2}\right]}\right] \tag{3}$$

$$\overline{R_{A\delta}} = \frac{\sqrt{M}\eta\,\overline{C_{Ai}}}{\sinh\sqrt{M}\eta}\frac{[R_2/(R_1\varphi_2)]}{1 + [R_2/(R_1\varphi_2)]} \tag{4}$$

$$\eta^2 = 1 + \frac{t_2}{q_0\left[1 + t_2 - \frac{1}{\cosh^2\sqrt{M}\eta\left\{1 + \frac{R_2}{R_1\varphi_3}\right\}}\right]}\left[1 - \varphi_2 - \frac{\left[1 - \frac{\varphi_2}{\cosh^2\sqrt{M}\eta}\right]}{\cosh\sqrt{M}\eta\left[1 + \frac{R_2}{R_1\varphi_2}\right]}\right] \tag{5}$$

, $\sqrt{M}\eta$, k_GH_A R_2 q_0 D 1 D 1 D 1 M $k_2D_AC_{Bb}$

where $\varphi_2=\frac{\text{[illegible]}}{\tanh\sqrt{M}\,\eta}$; $t_2=\frac{\text{[illegible]}}{k_L\varphi_2}=\frac{\text{[illegible]}}{R_G\varphi_2}=\frac{\text{[illegible]}}{p\varphi_2}$; $R_G=\frac{\text{[illegible]}}{k_G H_A a}$; $R_2=\frac{\text{[illegible]}}{k_L a}$; $R_1=\frac{\text{[illegible]}}{(1-a\delta)k_2 C_{Bb}}$; $M=\frac{\text{[illegible]}}{k_L^2}$;

$$q_0=\frac{D_B C_{Bb} H_A}{bD_A p_A}; \text{ and } p=\frac{D_B C_{Bb} k_L}{bD_A k_g p_A}; \tag{6}$$

Equation for η in the case of regime 2-3-4-5:

$$p\eta^3+\frac{q_0 \tanh\sqrt{M}\eta}{\sqrt{M}}\eta^2+(1-p)\eta-\frac{(1+q_0)}{\sqrt{M}}\tanh\sqrt{M}\eta=0 \tag{7}$$

Equation for η in the case of regime 3-4-5:

$$p\eta^3+\frac{q_0}{\sqrt{M}}\eta^2+(1-p)\eta-\frac{(1+q_0)}{\sqrt{M}}=0 \tag{8}$$

Equation for $\overline{R_A}$ in the case of regime 3-4-5:

$$\overline{R_A}=\frac{R_2}{R_G+\frac{R_2}{\sqrt{M}\eta}} \tag{9}$$

$$\frac{q_0}{\sqrt{M}}+p>>1 \tag{10}$$

$$\overline{R_A}=\frac{R_2}{[R_G+\frac{R_2}{\sqrt{M}}]} \tag{11}$$

$$\frac{q_0+1}{1-p}>3 \tag{12}$$

$$\sqrt{M}>30 \tag{13}$$

$$p<1 \tag{14}$$

$$\overline{R_A}=\frac{(q_0+1)}{(1+\frac{p}{q_0})} \tag{15}$$

$$\frac{q_0 \tanh\sqrt{M}}{\sqrt{M}}+\frac{p}{[\frac{\sqrt{M}\,R_2}{R_1 \tanh\sqrt{M}}+1]}>>1 \tag{16}$$

$$\overline{C_{Ai}}=\frac{\frac{p_A}{H_A}}{\frac{P\sqrt{M}}{q_0 \tanh\sqrt{M}}+1} \tag{17}$$

$$\overline{R_A}=\frac{\sqrt{M}\,\overline{C_{Ai}}}{\tanh\sqrt{M}} \tag{18}$$

$$\overline{R_{A\delta}}=\frac{\sqrt{M}\,\overline{C_{Ai}}}{\sinh\sqrt{M}} \tag{19}$$

$$\overline{R_A} = \frac{R_2}{R_2 + R_G + R_1^{'}} \tag{20}$$

where the parameter, $R_1^{'}$ is defined by the following equation:

$$R_1^{'} = \frac{1}{lk_2 C_{Bb}} = \frac{R_2}{M + R_2/R_1} \tag{21}$$

$$\overline{R_A} = \frac{q_0}{(q_0 + p)} \tag{22}$$

$$\overline{R_A} = \frac{R_2}{R_1^{\prime}} \tag{23}$$

$$\overline{R_A} = \frac{q_0}{p} \tag{24}$$

S. No.	Regime	criteria	Rate equations
1	1-2-3-4-5	$0.3 < \sqrt{M} < 3$; $\frac{q_0 \tanh\sqrt{M}}{\sqrt{M}} + \frac{p}{[\frac{\sqrt{M}\,R_2}{R_1 \tanh\sqrt{M}} + 1]} \ngtr\ngtr 1$; $\frac{q_0}{p} \ngtr\ngtr 1$	(1)-(6)
2	2-3-4-5	$\sqrt{M}n < 3$ and $[\frac{R_2}{R_1}\cosh\sqrt{M}\eta\ ] \gg 1$ Conservative criterion: $\frac{R_2}{R_1} \gg 1$	$C_{Ai} = \frac{\left(\frac{p_A}{H_A}\right)t_2}{1+t_2}$; $C_{Ab} = 0$; (3),(4)
3	3-4-5	$\sqrt{M}n > 3$	(8) , (9)
4	3-5	$\sqrt{M} > 3$, (10)	(11)
5	4-5	(12)-(14)	(15)
6	1-2-3-5	$0.3 < \sqrt{M} < 3$; (16)	(1)-(4) $\eta = 1$
7	2-3-5	$\frac{q_0 \tanh\sqrt{M}}{\sqrt{M}} + p \gg 1$, $0.3 < \sqrt{M} < 3$; $\frac{R_2}{R_1} \gg 1$	(17)-(19)
8	1-2-5	$\sqrt{M} < 0.3$; $R_2 \ngtr\ngtr R_1$; $R_2 \ngtr\ngtr R_G$;	(20)-(21)
9	2-5	$\sqrt{M} < 0.3$; $\frac{R_2}{R_1} \gg 1$; $R_2 \ngtr\ngtr R_G$	(22)
10	1	$\sqrt{M} < 0.3$; $R_2 \gg R_G$; $\gg 1$; $\frac{R_2}{R_1}$	(23)
11	5	[illegible]	(24)

11	5	p≥1, $\frac{40}{p\sqrt{M}}$>>1	(24)

376g Development of a Visual Library for Chemical Process Safety Education: Introduction to Sketchnoting

Monica Lamm, Angela Manno, Victoria Meeks, Verena Paepcke-Hjeltness, Amarjargal Mendee, Ann Gansemer-Topf

The ability to acquire visual thinking competency offers students a powerful and unique opportunity to engage with the discourse of their discipline (Rau, 2017). Many STEM disciplines and specifically chemical engineering relies on the ability interpret knowledge visually. Although engineering students are exposed to countless figures and diagrams in textbooks, lectures, and labs (Kelly & Akaygun, 2019), and rely on these representations to make meaning of course content, they are not explicitly taught how to develop visualizations on their own (diSessa, 2004). This creates a gap, termed a 'representation dilemma' (Rau, 2017), between understanding a visual and using that visual as an aid for learning new content. One way to overcome this problem is to develop students' visual representation competency (Dickmann et al., 2019). We are undertaking a project to address the representation dilemma problem by providing chemical engineering students with explicit instruction on sketchnoting, a method for visual thinking and notetaking, within the context of a core chemical engineering course, Chemical Process Safety.

This poster will describe our implementation of sketchnoting in a chemical process safety course, based on an implementation strategy for enhancing visual thinking skills in STEM courses (Gansemer-Topf, Paepcke-Hjeltness, Russell & Schiltz, 2021). The poster will include a description of the steps taken to develop a visual library of commonly used course terms (e.g., hazard, incident, pressure, people) and notetaking flags (e.g., equation, revisit, checkpoint), and the approach taken to embed explicit sketchnoting instruction sessions into the course.

References cited

Dickmann, T., Opfermann, M., Dammann, E., Lang, M. & Rumann, S. (2019). What you see is what you learn? The role of visual model comprehension for academic success in chemistry. Chemistry Education Research and Practice. 20, 804-820.

diSessa, A. A. (2004). Metarepresentation: Native competence and targets for instruction. Cognition and Instruction, 22(3), 293-331.

Gansemer-Topf, A.M., Paepcke-Hjeltness, V., Russell, A.E. & Schiltz J. "Drawing" your Own Conclusions: Sketchnoting as a Pedagogical Tool for Teaching Ecology. Innov High Educ 46, 303–319 (2021).

Kelly, R. & Akaygun S. (2019). Chemistry Education Research and Practice 20, 657-658.

376h Artificial Intelligence (AI) Assisted Thermodynamics Learning

Karen Agro, David DiBiasio, Michael Timko

In this moment of proliferation of artificial intelligence (AI), major questions exist as to how to best approach these advancing technologies from a pedagogical standpoint. Ignoring them entirely has the potential to leave students at a disadvantage for their future work and careers, however, unrestricted use has the potential for issues of academic integrity. This study takes the approach that there can be a middle ground in which the student drives the AI in order to allow them to think more about the overall problem at hand, rather than the granularities of writing complex code. This is particularly evident in the area of thermodynamics.

Within thermodynamics, a limiting factor in solving certain complex problems is the student's ability to carry out tedious mathematical calculations in addition to their engineering analysis. This study utilizes ChatGPT to assist students in writing MATLAB code without needing an independent and extensive command of MATLAB coding language. ChatGPT working in collaboration with the students can produce the general structure of the required code for the student to then customize based on their engineering analysis and specific thermodynamics question. This allows us to imagine a future where the focus is on problem set up, rationale, critical thinking, and engineering analysis, without being limited by tedious calculations. This teaching methodology was rolled out in a graduate level thermodynamics course utilizing ChatGPT to write MATLAB code for systems of ordinary differential equations, molar volume and fugacity calculations of pure gases and binary gas mixtures, vapor liquid equilibrium calculations, etc. Students were surveyed both before and after taking thermodynamics with AI to assist in coding.

376i Investigating Students' Experience with Generative Chatbots in Chemical Engineering Education

Fiammetta Caccavale, Carina L. Gargalo, Krist V. Gernaey, Ulrich Krühne

Artificial Intelligence (AI), mainly thanks to the rise of Large Language Models, has recently created much traction, forcing educators worldwide to reflect upon its potential use (and misuse) if used in coursework. These systems have gained considerable interest in educational research and practice due to their ability to learn from and generate natural language. Thanks to the extensive pre-training these models are exposed to, they obtain knowledge about a wide range of topics and are suitable for application in many different contexts and fields. This ability to extract content and critical information from text offers a powerful tool for enhancing the learning experience. According to Huang et al. (2022), these AI tools herald a new era in education, offering unparalleled potential to revolutionize learning experiences and outcomes. They could be used to identify knowledge gaps, aid formative assessment, and provide personalized feedback (Hopfenbeck, 2023; Dai et al., 2023). Researchers have also been evaluating how LLMs perform in a real educational context. Hicke et al. (2023) assess the efficacy of LLMs in generating accurate teacher responses. Xiao et al. (2023) apply LLMs to real-world classroom settings, implementing a reading comprehension exercise generation system that provides students with high-quality and personalized reading materials.

This work focuses on intelligent tutoring systems in a real educational context. We introduced a software tool called ChatGMP, an LLM trained in Good Manufacturing Practice (GMP) data, to a course taught at the Technical University of Denmark (DTU). In this course, students are required to participate in an audit exercise, where they pretend to be an auditor inquiring a company, represented by the teachers, about their good manufacturing practices. During the audit, students can ask questions regarding available documents and practices adopted by the company, which are indispensable to evaluate whether the company in question could be a valuable business partner or if the non-conformities found are too severe.

This year, we started investigating the replacement of the physical teachers in this exercise with a generative AI model, called *ChatGMP*. The model is a pre-trained Transformer (Vaswani et al., 2017), FLAN-T5 (Chung et al., 2022), used to perform a question-answering task. The answers are retrieved through prompt engineering, where semantic search is performed over the historical database of question-answer pairs. To enable the students to interact with the intelligent tutoring system, which was presented in the form of a chatbot, we built an interactive GUI. *ChatGMP* was tested in three audits out of twenty-one total groups. The model managed to complete the three audits, answering the students' questions in a meaningful and complete manner and showing all the requested documents. A post-audit survey shows that all three groups participating in the audit exercise were happy about the experience and were satisfied with the quality of the responses. This marks the first step in successfully automating this task, which would enable more students to attend the course, since currently the yearly uptake is limited by the extensive time required by this exercise.

References

Huang, K. F. Hew, L. K. Fryer. (2022). Chatbots for language learning—are they really useful? a systematic review of chatbot-supported language learning, Journal of Computer Assisted Learning 38. 237–257.

T. N. Hopfenbeck. (2023). The future of educational assessment: self-assessment, grit and chatgtp?

W. Dai, J. Lin, F. Jin, T. Li, Y.-S. Tsai, D. Gasevic, G. Chen. (2023). Can large language models provide feedback to students? a case study on chatgpt.

Y. Hicke, A. Masand, W. Guo, T. Gangavarapu. (2023). Assessing the efficacy of large language models in generating accurate teacher responses, in Proceedings of the 18th Workshop on Innovative Use of NLP for Building Educational Applications (BEA 2023), Association for Computational Linguistics, Toronto, Canada, pp. 745–755. URL: https://aclant hology.org/2023.bea-1.60. doi:10.18653/v1/2023.bea-1.60.

Xiao, S. X. Xu, K. Zhang, Y. Wang, L. Xia. (2023). Evaluating reading comprehension exercises generated by LLMs: A showcase of ChatGPT in education applications, in Proceedings of the 18th Workshop on Innovative Use of NLP for Building Educational Applications (BEA 2023), Association for Computational Linguistics, Toronto, Canada, pp. 610–625. URL: https://aclanthology.org/2023.bea- 1.52. doi:10.18653/v1/2023.bea-1.52.

Vaswani, A., Shazeer, N., Parmar, N., Uszkoreit, J., Jones, L., Gomez, A. N., ... & Polosukhin, I. (2017). Attention is all you need. Advances in neural information processing systems, 30.

Chung, H. W., Hou, L., Longpre, S., Zoph, B., Tay, Y., Fedus, W., ... & Wei, J. (2022). Scaling instruction-finetuned language models. arXiv preprint arXiv:2210.11416.

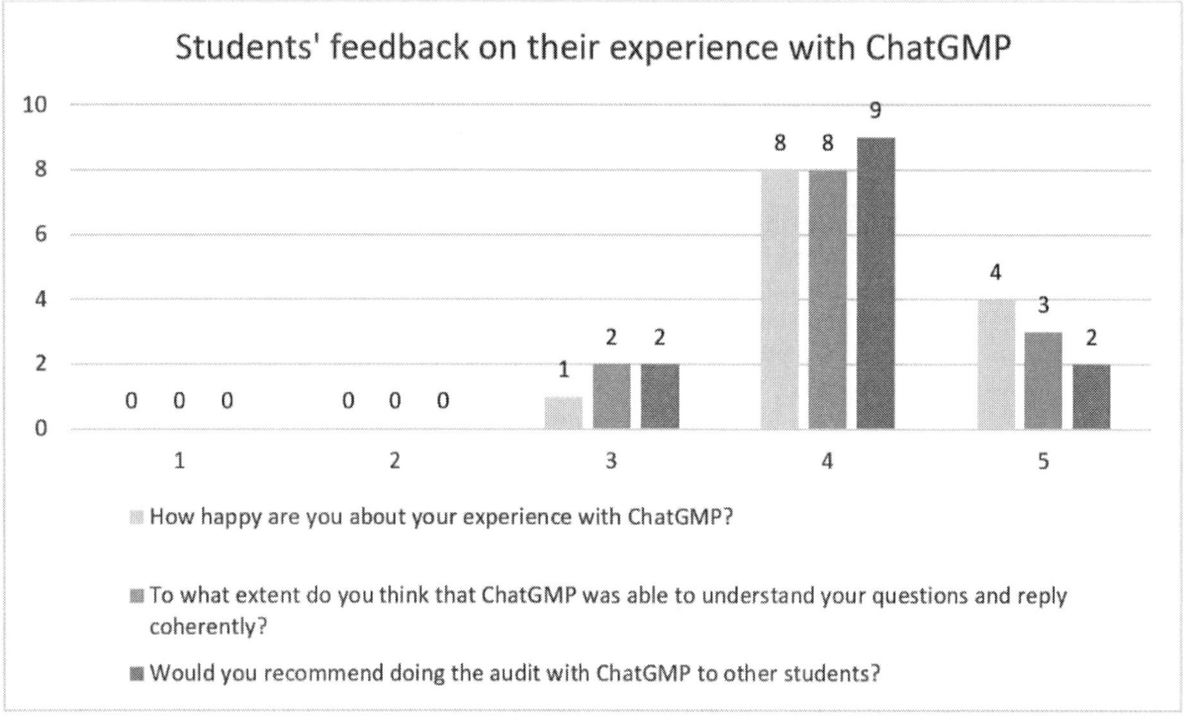

376j Assessing Team Performance in Control Rooms: Integrating Eye Tracking with Multimodal Data Analysis for Insights

Haider Altaf, Babji Srinivasan, Rajagopalan Srinivasan

The Control Room is like the brain of the system, where multiple operators play their role as a team to ensure the plant operates safely and efficiently in both normal and emergencies. As automation and AI tools become increasingly prevalent, operators face handling a large volume of data (Naef and Lefsrud, 2023). This growing data provides opportunities for deeper insights. However, it simultaneously poses challenges, including the risk of overwhelming operators with information overload, which can affect their judgment and decision-making capabilities. In such situations, effective communication and collaboration among operators become crucial for maintaining situational awareness and making informed decisions (Kluge et al., 2014). Shared Mental Model, defined as "organized knowledge shared among team members" by Orasanu & Salas(1993), plays a crucial role in fostering common understanding and interpretation of tasks, leading to improved teamwork, coordination, and communication (Kluge et al., 2014). Teams with a strong Shared Mental Model demonstrate faster reactions, greater accuracy, and higher mission success rates, while also enhancing their capacity to handle novel tasks and situations (Espevik et al., 2011). This collective understanding and alignment among team members contribute significantly to overall team performance and effectiveness.

To evaluate Shared Mental Model, it's crucial to delve into the cognitive behaviors of team members and understand their situational awareness. Recent technological advancements have enabled physiological measures like eye movement to study cognitive behaviors of the operators in modern Control Rooms. For instance, Salehi et al. (2018) found that novice operators generally focused less (as evidenced by shorter fixation durations) on pivotal HMI (Human-Machine Interface) areas compared to their expert counterparts. In our previous work, we have delved deep into cognitive processes and mental models using eye tracking technology. Our findings reveal that operators familiar with process dynamics focused primarily on essential variables pertinent to the current situation (Bhavsar et al., 2017).

Additionally, we developed a novel method using eye tracking to measure operators' attention in control rooms with multiple displays, including Large Screen Displays (LSD) for shared situational awareness. Alongside cognitive aspects, assessing team communication is integral to understanding Shared Mental Model. Study by Kluge et al. (2014) revealed that teams with Shared Mental Models engaged in less explicit communication, relied more on information transfer. Another study by Khawaja et al. (2012) showed that teams exhibited decreased agreement and increased disagreement during challenging tasks due to high cognitive load. Given these insights, there is a pressing need to devise a comprehensive methodology that utilizes multimodal data sources, such as operator attention, interaction patterns, and actions, to gain a holistic understanding of operator behaviors conducive to efficient teamwork.

In this work, we proposed a novel methodology for evaluating team performance in collaborative process control rooms, with a particular focus on environments featuring Large Screen Displays (LSD) positioned separately from individual workstations, a unique aspect not previously explored. Our approach utilizes eye-tracking technology to measure attention within the control room setup. Additionally, we incorporate multimodal data sources, including voice and text messages, to analyze team interactions, along with process information such as operator actions, alarm responses, and task completion times, providing further insights. The proposed method is validated on an in-house test bed replicating the environment of a collaborative control room with multiple displays, including one LSD tailored for collaborative monitoring. Human subjects, acting as operators in an Ethanol production unit simulator control room, participated in the validation process. The results highlighted the effectiveness of our methodology in capturing team performance in modern collaborative control room setups. The collected data were analyzed to measure the situational awareness, important information shared, and finally, action taken to measure team performance, shedding light on shared attention patterns, team dynamics, and collaborative decision-making processes. These insights are pivotal for enhancing individual and team performance and understanding Shared Mental Models across various industrial settings.

References:

Naef, M., & Lefsrud, L. (2023). Smooth operator: Aligning performance assessment methods with design and operating objectives. *Journal of Loss Prevention in the Process Industries*, 105158.

Kluge, A., Nazir, S., & Manca, D. (2014). Advanced applications in process control and training needs of field and control room operators. *IIE Transactions on Occupational Ergonomics and Human Factors*, *2*(3-4), 121-136.

Orasanu, J., & Salas, E. (1993). Team decision making in complex environments.

Espevik, R., Johnsen, B. H., & Eid, J. (2011). Communication and performance in co-located and distributed teams: an issue of shared mental models of team members?. *Military Psychology*, *23*(6), 616-638.

Salehi, S., Kiran, R., Jeon, J., Kang, Z., Cokely, E. T., & Ybarra, V. (2018). Developing a cross-disciplinary, scenario-based training approach integrated with eye tracking data collection to enhance situational awareness in offshore oil and gas operations. *Journal of Loss Prevention in the Process Industries*, *56*, 78-94.

Bhavsar, P., Srinivasan, B., & Srinivasan, R. (2017). Quantifying situation awareness of control room operators using eye-gaze behavior. *Computers & chemical engineering*, *106*, 191-201.

Khawaja, M. A., Chen, F., & Marcus, N. (2012). Analysis of collaborative communication for linguistic cues of cognitive load. *Human factors*, *54*(4), 518-529.

376k Using an Industrial Example to Teach Pollution Prevention and Residence Time Distributions

Robert Hesketh, Kirti Yenkie, Barnabas Gao, Steven Roth, David Theuma, Emma Padros, C. Stewart Slater, Mariano J. Savelski, Michael Fracchiolla

A practical example will be presented that can be used to teach students the mathematics of residence time distribution theory as well as pollution prevention. Undergraduate chemical engineering students have the opportunity to learn about residence time distributions in chemical reaction engineering, but only if they get to chapters 16 -18 of Fogler's Elements of Chemical Reaction Engineering. But if professors follow the guidance by Fogler in the section titled, "Who is the intended audience" then they will never be taught about residence distribution theory since it is recommended only for graduate students. In addition if you talk with the chemical engineering professor sitting next to you they will probably give you one of these reasons that they can not teach residence time distribution theory because it is not practical; too difficult to teach; they don't understand it well enough to teach it; or they never saw it used in practice. But what we are seeing in industry is the use of residence distribution times in the pharmaceutical industry in the area of continuous processing. This is an area that undergraduate chemical engineers have been hired to work in by major pharmaceutical industries. So shouldn't they have some knowledge about this subject? The other question that students ask is are all reactors ideal? Do we need to know about models other than perfect mixing or plug flow to describe processing equipment? In addition to residence time distributions we are also recognizing that we need more examples of how chemical engineers can contribute to the Sustainability objectives of major chemical companies. This presentation will show an example of pollution prevention which can be used to help meet this objective.

3761 Social Justice and Societal Impact of Developing a Green Methanol Process in West Virginia

Kaleb Schifflett, Srinivas Palanki, Xingbo Liu

West Virginians bear much of the health risks caused by natural gas, mining, and chemical process industry in the Appalachian region. This mix of pollutants includes significant amounts of greenhouse gases. Toxic air pollutants cause cancer and other diseases, while ozone smog can cause a variety of health problems, such as asthma attacks and worsening the effects of bronchitis and emphysema. Approximately 46% of West Virginia's population lives in disadvantaged tracts. In this talk, we will focus on the production of methanol and the social justice implications of developing a novel methanol process that significantly reduces the utilization of methane and the emission of greenhouse gases. The focus of the energy and environmental assessment will be in counties that have disadvantaged tracts in terms of air quality and climate change.

Methanol is a colorless, flammable liquid that is used as a solvent, fuel, and feedstock in the production of a wide variety of chemicals and materials. It has traditionally been produced from natural gas by first converting methane to syn gas and then converting syn gas to methanol. However, this is a very energy intensive process and produces a significant amount of the greenhouse gas carbon dioxide. In this research a novel conceptual design of an integrated process for producing green methanol is developed in the ASPENPlus simulation environment. This data is analyzed in terms of the carbon dioxide removed, energy utilized, and potential waste material generated and utilized in the Department of Energy's Justice 40 assessment using the Environmental Protection Agency's (EPA) screening tool EJScreen. This tool is used to determine demographic socioeconomic and environmental information for specific areas that have a large concentration of chemical and power industry, which potentially generate the most amount of carbon dioxide. Environmental and demographic indicators are combined to determine EJ indexes. We utilize 12 environmental indicators and 7 socioeconomic indicators to generate 12 EJ indexes that are used for the preliminary energy and environmental justice assessment. The color-coded mapping and the standard report for each selected area are compared to the state, EPA region, and the nation in the assessment. We discuss how these results are disseminated in our undergraduate and graduate chemical engineering curriculum.

Acknowledgment

This study was supported by the United States Department of Energy

376m Fostering Holistic Understanding of Sustainability through a Mixed-Methods Study in Foundry-Guided Courses

Shafieh Karami, Andrea Arce-Trigatti, Pedro Arce

The present study explores how graduate students' comprehension of sustainability principles evolve through their engagement in courses guided by the Renaissance Foundry Model (i.e., Foundry) – an innovation-driven learning platform. Investigating this topic is highly relevant for Chemical Engineering and related STEM curricula, as it aligns well with the theme of fostering sustainability and global awareness in the classroom as a foundation to develop holistic-style professionals. In accordance, scholars and leading professional organizations, including the National Academy of Engineering (NAE) (2013, 2023), Lemelson Foundation (2023), and the National Science Foundation (NSF) (2023) emphasize the importance of incorporating sustainability principles and design thinking approaches into engineering education to nurture future professionals capable of addressing intricate global challenges. According to the NSF (2023) "The complex problems facing society in the 21st Century, including the impact of the COVID-19 pandemic, demand changes to the way engineers are educated and the integration of new modes of learning for engineering students. For example, solving the National Academy of Engineering (NAE) Grand Challenges will require engineers who not only have deep technical knowledge, but also an understanding of the societal and global contexts in which those problems occur." To this end, incorporating key concepts like interdisciplinary collaboration, innovation-driven learning, critical thinking, prototype development, and sustainability principles into engineering curricula are essential components to the development of this type of professional.

Further, with respect to sustainability, the Lemelson Foundation (2023) has made significant efforts in the development of the Engineering for One Planet (EOP) framework which offers a taxonomy of nine principles relevant to understanding different layers of sustainability in design. These principles include, for example, systems thinking, social responsibility, materials selection, critical thinking, teamwork and communication, business responsibility, and environmental literacy, which are not only relevant to chemical engineering concepts, but emphasize the design thinking aspects of the discipline. In tandem with an innovation-driven learning platform, like the Foundry, students are responsible for developing a prototype of innovative technology that addresses a community-centered challenge with a real-world application that incorporates sustainability with design thinking processes and community collaboration (Wagale et al., 2023, 2024). Thus, wherein the EOP framework emphasizes the integration of sustainability principles across various disciplines, the Foundry provides students with hands-on, project-based learning experiences through community immersion trips that foster design thinking, community engagement, and innovation (Pabody et al., 2023).

In this IRB approved research, we propose a mixed-methods study that examines graduate students' understanding of sustainability before and after participating in Foundry-guided courses related to an NSF-NRT program at Tennessee Tech University. Two courses offer the context for this study: Introduction to Food-Energy-Water (FEW) Nexus and FEW Nexus Challenges, both Foundry-guided courses with immersion experiences, with the second course emphasizing the EOP framework in design processes. These courses provide students with opportunities to apply these concepts to real-world challenges, fostering a holistic understanding of sustainable practices and innovative problem-solving by developing Prototype of Innovate Technology (PIT). The data collected will encompass survey responses assessing pre- and post-course understanding, as well as reflective entries from students' personal class binders (PCBs) throughout the course (Rawlings et al., 2005). The primary analysis for this study will involve qualitative evaluation of students' PCB entries and quantitative analysis of survey responses, allowing for a comprehensive assessment of students' growth in understanding sustainability aspects. Relevant implications from this study will highlight effective pedagogical strategies for integrating sustainability into chemical engineering curricula, ultimately contributing to the development of globally conscious and responsibly innovative engineers. Future steps related to this research include longitudinal studies tracking students' retention and application of sustainability concepts in their professional endeavors, as well as the potential expansion of Foundry-guided courses across various engineering disciplines.

448a How We Teach Teaching: Graduate Student Instructor Training in Chemical Engineering

Christopher Chen, Lisa Bullard, Adam Melvin

In Spring 2024, the Education Division surveyed departments in the US and Canada over how they trained or otherwise prepared graduate student instructors (GSIs) and the teaching activities of GSIs in these roles. We use the term graduate student instructors to denote a very broad range of teaching positions that graduate students may hold in the department including, but not limited to, instructors, teaching assistants (TAs), graders, readers, and lab managers, as all of these positions support the learning of chemical engineering students. This survey has reached departments big and small (from those with 3 GSIs to over 100); those with and without teaching requirements for their graduate students (0 to 2+ years of required teaching); and a wide range of training hours for GSIs (from 0 hours of training to 8+ hours of orientations, workshops, and discipline-specific pedagogy sessions).

In this first-of-its-kind survey on engineering GSI training (as far as the authors are aware), the goal of the project is to help departments better understand and learn from others how they can improve their development of GSIs both as essential instructors for current students, but also to prepare graduate students as future faculty in chemical engineering. Trends in topics for discipline/department-specific training will be shared, along with how we assess and give feedback to GSIs on their teaching. Suggestions as to how ASEE/AIChE can help to support GSI training initiatives will also be shared and discussed.

448b How We Teach: Transport Phenomena and Applications

Laura Ford, Janie Brennan, Jacqueline Burgher Gartner, Matthew Cooper, Kevin Dahm, Luke Landherr, David Silverstein, Stephen Thiel, Troy Vogel

The Education Division will survey departments in the US and Canada over the teaching of transport phenomena and applications in spring and fall of 2024. Equilibrium-based separations will be included. The spring survey asks only about the number of courses within the area, their topics, and a contact faculty member. The fall survey will follow up with a customized survey for each faculty member. Preliminary results have been received from 39 institutions. Departments most commonly have 3 or 4 courses in this broad area (38% of respondents each). Having fewer courses is more common than more than 3 or 4: 8% have 1 course, 10% have 2, and 5% have 5 or more. More details will be presented at the meeting after the second survey has been completed.

448c Future Directions in Teaching Transport Phenomena and Applications

Arthur Hersel

To prime the pump for the discussion period, Prof. (and textbook author) Hersel will present his thoughts on three topics related to the teaching of transport phenomena and applications. 1) Should topics like heat transfer include design exercises, or is it more appropriate to save such activities for later stages of learning? 2) With tools like Wolfram Alpha readily available, is there still a need for students to focus on mathematical skills? Should we integrate computational methods such as COMSOL into the curriculum? 3) What is the proper balance between multi-scale modeling and analysis in textbooks and course content?

197a Student Identity and Belonging As It Relates to Gender and Gpa

Stephanie Velegol, Mechteld Hillsley, Karen High, Kate Getz, Jennifer Brown, Terri Vescio

The purpose of this presentation is to explore the relationship between GPA and engineering identity and belonging for chemical engineering students across genders. We surveyed a total of over 300 chemical engineering students at Penn State ranging from the 2nd year to 4th year. The survey included questions from four validated instruments:

1. Identity: Engineering Identity (Godwin 2016)
2. Identity: Collective Self-Esteem (Luhtanen and Crocker 1992)
3. Belonging: Sense of Social and Academic Fit (Walton *et al.* 2023)
4. Belonging: Peer Inclusivity (Davis *et al.* 2023)

There is no correlation between GPA and gender however we see significant differences with gender for all four instruments above. We further explored this data to look for correlations between students' GPA and how these correlations might be different for students who identified as male and those who identified as female. We found that GPA did not correlate with any of the instruments for those identifying as men but did correlate with Engineering Identity and Sense of Social and Academic fit for those identifying as women. We will discuss these differences in terms of other influences including growth and fixed mindset.

197b Promoting Chemical Engineering Identity through Student Agency and Experiment Relevance

Eva Chi, Madalyn Wilson-Fetrow, Catherine Hubka, Jennifer Brown, Stephanie Wettstein, Vanessa Svihla

Problem

Laboratory courses play a crucial role in the education of chemical engineers. However, early chemistry courses often focus on cookbook style laboratory experiments, which can shape students' expectations about upper division chemical engineering laboratory courses 1, 2. In such courses, students are often highly scaffolded in all aspects of their work, leaving them little room to make meaningful decisions. Furthermore, considering the foundational chemistry courses and chemical engineering laboratory courses, most experiments are predetermined, meaning the solution is known ahead of time. While such laboratory experiments are easier to instruct, they offer students limited opportunities for decision-making; in other words, the students have little agency.

ABET criteria have made clear, however, that students should have agency: "an ability to develop and conduct appropriate experimentation, analyze and interpret data, and use engineering judgment to draw conclusions" 3. In particular, the idea that students should "develop"—not just conduct—and "use engineering judgment" suggests a level of agency that far exceeds what students experience in the typical introductory chemistry laboratories. In response, faculty in chemical engineering have studied varied higher agency approaches 4, including course-based undergraduate research experiences 5, 6, integrating data science 7, and enhancing student agency over specific aspects, like making choices about data analysis in a unit ops laboratory 8.

Our strategy to enhance student engagement and sense of belonging in chemical engineering involves examining the influence of consequential agency in laboratory experiments. We aim to assess the impact of agency on students by investigating various aspects of laboratory experiments through a survey measuring consequential agency, engineering identity, relevance, and persistence intentions. The project aims to expand existing theories on agency in learning and the formation of professional identity in advanced chemical engineering laboratory courses.

To assess the influence of making consequential decisions on student development, we investigated students' perspectives on four domains within laboratory experiments: (1) experimental design prior to doing the laboratory experiment; (2) data collection and documentation during experiments; (3) data analysis and interpretation; and (4) communication of purpose, methods, and conclusions—contributes to students' development. In addition, we wondered about the impact of perceived relevance—that is, the degree to which students viewed the experiments as similar to or relevant to the work professional chemical engineers do. In terms of student development, we investigated students' persistence intentions and engineering identity. We posed a research question:

- To what extent do study constructs (consequential agency in four domains, relevance) explain variance in students' identities as engineers?

We hypothesized that having agency in domains like 1 and 3 may matter more than domain 2.

Methods

We used a previously developed survey to assess consequential agency across each of the four domains, as well as engineering identity and relevance. A total of 74 students from two research universities participated in the survey as part of their post-lab activities. At one university, seniors were enrolled in a 3-credit laboratory course (the 2nd in a 2-semester sequence) in which they completed four-unit operations experiments (a heat exchanger experiment, fluid flow and friction, a continuous stirred tank reactor (CSTR) experiment, and an enzyme reaction kinetics experiment). At the other, juniors were enrolled in a

1-credit laboratory course (the 2nd in a 4-semester sequence) in which they completed three transport phenomenon experiments (investigating how pipe length impacts efflux from a tank, how types of wood impact conductive heat transfer, and how fittings and pipes impact frictional pressure loss); seniors were enrolled in a 1-credit laboratory course (the 4th in a 4-semester sequence) in which they completed two kinetics and process control experiments (investigating different catalysts used in the selective hydrogenation of ethane, and determining and comparing tuning parameters for a level controller on a water tank).

We used regression modeling in order to investigate trends in the data.

Results

We found engineering identity was positively and significantly predicted by perceived relevance and data analysis and interpretation consequentiality (Domain 3), and negatively, significantly predicted by consequentiality during the experiment (Domain 2), $F(58, 5) = 4.03$, $r2 = .129$ (Table 1). Neither consequential agency in Domains 1 nor 4 explained significant variance.

Implications

This study examined how having consequential agency in each domain, alongside perceptions of relevance, influences student development. Our results showed that students exercising agency in tasks such as data analysis and result interpretation (Domain 3) contributes more to their identity as engineers compared to other domains. Similarly, tasks perceived as relevant, meaning that students believe they resemble the work of chemical engineers, offer a more impactful experience for students' evolving engineering identities. Although our results are preliminary, they suggest that faculty aiming to enhance their laboratory courses can prioritize empowering students to make decisions regarding analysis and interpretation, while also helping them grasp the connections between experiments and the work of chemical engineers. However, it's important to acknowledge a notable limitation: relatively few students reported high agency experiences in Domain 1. Consequently, our findings indicating that consequentiality in Domain 1 did not account for variance in engineering identity may be attributed to the limited variability in scores on Domain 1. Given that this domain is where students are least likely to exercise agency9, 10, our ongoing research endeavors are strategically designed to identify such instances to ensure comprehensive modeling of the impacts of each domain.

References

(1) Kirn, A.; Benson, L. Engineering students' perceptions of problem solving and their future. *Journal of Engineering Education* **2018**, *107* (1), 87-112. DOI: 10.1002/jee.20190.

(2) Pisani, S.; Haw, M. D. Learner agency in a chemical engineering curriculum: Perceptions and critical thinking. *Education for Chemical Engineers* **2023**, *44*, 200-215. DOI: 10.1016/j.ece.2023.06.003.

(3) ABET Inc. *Criteria for Accrediting Engineering Programs, 2024 – 2025.* 2024. https://www.abet.org/accreditation/accreditation-criteria/criteria-for-accrediting-engineering-programs-2024-2025/ (accessed 1/2024).

(4) Anagnos, T.; Komives, C.; Mourtos, N. J.; McMullin, K. M. Evaluating student mastery of design of experiment. *Proceedings of the Frontiers in Education Conference* **2007**, T3G-7-T3G-12. DOI: 10.1109/FIE.2007.4417923.

(5) Wilson-Fetrow, M.; Svihla, V.; Burnside, B.; Datye, A. K. Course-based undergraduate research experiences in a chemical engineering laboratory promote consequential agency. *Journal of Chemical Education* **2023**, *100* (10), 3752-3763. DOI: 10.1021/acs.jchemed.2c00582.

(6) Giordano, A. N.; Styers-Barnett, D. Training tomorrow's scientists: Embedding professional skills in the physical chemistry curriculum with a guided grant and laboratory project. *Journal of Chemical*

Education **2022**, *99* (6), 2417-2424. DOI: 10.1021/acs.jchemed.1c00817.

(7) Helmbrecht, H. Effective laboratory education with TEXTILE: Tutorials in EXperimentalisT Interactive LEarning. *Chemical Engineering Education* **2022**, *56* (4), 1-11. DOI: 10.18260/2-1-370.660-129820

(8) Elkhatat, A.; Al-Muhtaseb, S. A. Fostering Engineering Laboratory Course Teaching by Embedding an Inquiry-Guided Learning Approach Using Computer-Aided Learning Packages: Evaluation of Learning Outcomes in a Cooling Tower Experiment in the Unit Operations Lab. *Chemical Engineering Education* **2022**, *56* (3), 190-198. DOI: 10.18260/2-1-370.660-129133.

(9) Stoodley, R.; Knox, K. J.; Gillis, E. A. Application of a cognitive task framework to characterize opportunities for student preparation for research in the undergraduate chemistry laboratory. *Journal of Chemical Education* **2023**, *100* (5), 1831-1842. DOI: 10.1021/acs.jchemed.2c00948

(10) Wieman, C. Comparative cognitive task analyses of experimental science and instructional laboratory courses. *The Physics Teacher* **2015**, *53* (6), 349-351. DOI: 10.1119/1.4928349.

Table 1. Regression model of engineering identity

	Unstandardized Coefficients		Standardized Coefficients		
	B	Std. Error	Beta	t	Sig.
(Constant)	2.65	0.88		3.00	0.00
Relevance	0.32	0.14	0.29	2.35	0.02
Domain 1 Consequentiality	0.10	0.09	0.14	1.11	0.27
Domain 2 Consequentiality	-0.20	0.09	-0.29	-2.19	0.03
Domain 3 Consequentiality	0.21	0.11	0.26	1.87	0.07
Domain 4 Consequentiality	0.11	0.12	0.13	0.93	0.36

197c Leveraging Qualitative and Quantitative Methods to Transform Chemical Engineering Laboratories into Remote Learning Environments

Sindia M. Rivera-Jiménez, Fernando Mérida Figueróa

This study exemplifies active engagement in the scholarship of teaching and learning (SoTL) by presenting a practical methodology for enhancing remote chemical engineering laboratories through qualitative and quantitative approaches. Our primary objective is to address the challenge of leveraging student feedback for informed instructional design and curriculum development using evidence-based practices. We focus on answering the question: What are the most effective learning objectives and pedagogical approaches for integrating these desk-scale experiments into a blended learning environment in chemical engineering laboratories?

To explore the potential transformation of the Unit Operation 1 lab into a blended environment, we prioritize learning objectives that reinforce fundamental principles through hands-on experiments and that are not typically covered in lecture courses. From Fall 2020 to Fall 2023, students enrolled in Unit Operations 1 Laboratory courses at the University of Florida provided feedback through Likert-scale surveys and open-ended questions. We systematically analyzed this feedback using descriptive statistics (quantitative) and thematic analysis (qualitative).

Our findings offer valuable insights into student experiences and perceptions, showcasing the careful integration of quantitative and qualitative data to inform instructional design and curriculum development. Findings underscore the importance of combining theoretical concepts with practical, hands-on experiences to deepen student understanding and foster skills development. Recommendations derived from both qualitative and quantitative analyses offer practical guidance for educators seeking to design engaging and effective laboratory experiences. These recommendations include the design of interactive online modules, clear instructional materials, and ongoing incorporation of student feedback to refine curriculum delivery.

This study demonstrates a commitment to advancing SoTL within the chemical engineering education community by providing a structured approach to bridging the gap between theoretical knowledge and practical application in remote learning environments. By employing a combination of qualitative and quantitative methods, our approach empowers educators with the tools to effect meaningful classroom transformations. Qualitative insights from student feedback offer rich contextual understanding, complemented by the objective assessments of learning outcomes provided by quantitative analyses. This dual approach enables educators to tailor instructional methodologies effectively, ensuring a comprehensive educational experience for all students.

197d Turbocharging Classroom Sense of Belonging Using the 'Conocimiento Activity': Implementation Strategies, Assessment Results and Lesson's Learnt

Mandar Kathe, Benjamin Bouton

In any chemical engineering classroom, fostering a strong sense of belonging is paramount. This intangible yet powerful feeling significantly impacts students' academic experiences, well-being, and overall success. There are several classroom and longitudinal studies that show that an increased sense of belonging fuels intrinsic motivation, encouraging students to invest time and effort in their studies[a]. Sense of belonging is positively associated with academic performance, retention, and persistence, and it is particularly important for college students from marginalized groups[b]. Professors play a crucial role in creating a sense of belonging. Encouraging open communication, providing mentorship, and acknowledging students' diverse backgrounds contribute to a positive classroom atmosphere.

The process of Conocimiento[c] was developed in early 1970's. It is rooted in the inspiration from Paolo Freire, a Brazilian educator who believed that we could teach each other through dialogue. It is a basic tool for developing group connections. At Ohio State University, in senior chemical engineering classes, we use the process of Concimiento as a starting point to foster dialogue while also providing foundation that provides professors and students a reference point to build their scholarly community. The use of Conocimiento activity as a tool for improving a sense of belonging in classroom is driven by two basic paradigms:

- We as engineers depend on each other and the insights that we all contribute to better navigate Ohio State University, the college, and beyond. We also depend on each other to reach our professional goals. We must learn who we are so that we have a context for the individuals who make up our community of colleagues
- Sharing our stories and listening to others is essential for developing strong discussions.

In this talk, we will initially present a comprehensive literature search on various methods and assesment measures for sense of belonging in a classroom. We will then provide details of implementation of the 'Conocimiento activity'. Following the details, we will provide assessment measures and formative feedback results that were obtained regarding student sense of belonging. Specifically, we will provide a pre- and post- measure of class-room sense of belonging (obtained by adapting the PSSM tool[b]). We will present the adaptation of Python's NLTK library and Microsoft Azure tools for student sentiment analysis (as shown in the Figure attached). Finally we will provide 'what worked', 'lessons learnt' and 'reflections' for instructors that are interested in adapting this activity and working on improving a 'Sense of belonging' in their chemical engineering classroom.

References

a) Hansen, M.J., Palakal, M.J. & White, L. The Importance of STEM Sense of Belonging and Academic Hope in Enhancing Persistence for Low-Income, Underrepresented STEM Students. *Journal for STEM Educ Res* (2023). https://doi.org/10.1007/s41979-023-00096-8

b) Freeman, Tierra M., Lynley H. Anderman, and Jane M. Jensen. 'Sense of belonging in college freshmen at the classroom and campus levels.' *The Journal of Experimental Education* 75.3 (2007): 203-220.

c) Bridging and Breaking dialogues of belonging (https://belonging.berkeley.edu/sites/default/files/dialogues_of_belonging_-_bridging_and_breaking)

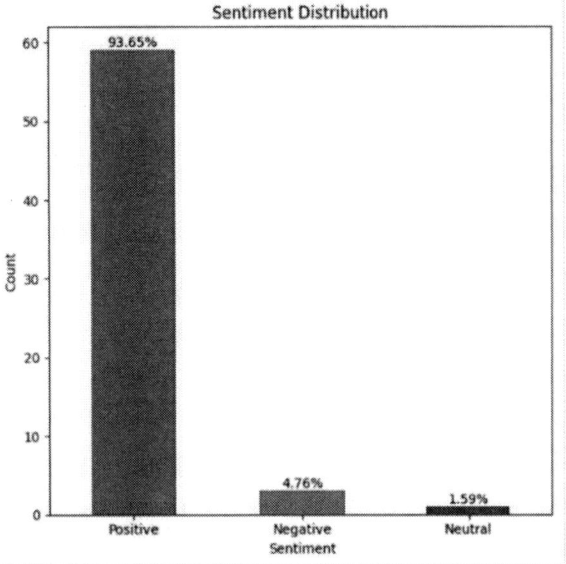

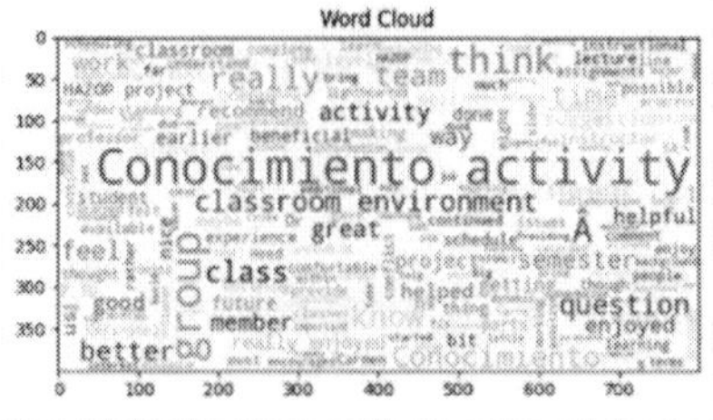

Example Limitations due to question formulation:

- This may not be possible, but if TAs could be present during group work time in class to serve as an additional point of contact for questions, it would dramatically reduce the lines that form to ask the instructor questions.Â The classroom is fine, no complaints there. I have not had our Conocimiento yet, as our schedules kept forcing our group to postpone. Sentiment tag: Negative
- Conocimiento was goodÂ Sentiment tag: Neutral

197g A Personalized Learning Model for STEM Graduate Education - a Pilot in Chemical Engineering

Susan Fullerton, Götz Veser, April A Dukes, Mary Besterfield-Sacre

Most U.S. graduate engineering programs traditionally follow a "one-size-fits-all" approach that focuses narrowly on research skills, is slow to incorporate industry trends, and defaults to training students for careers as university professors. Further, they implicitly assume that all students start at the same level of knowledge, disregarding differences not only in academic preparation, but in students' background, including socioeconomic, sociocultural, prior work experience and professional development. However, not accounting for these factors decreases inclusivity. Guided by the principles of personalized learning, and supported by an Innovations in Graduate Education grant from the National Science Foundation, the department of Chemical Engineering at the University of Pittsburgh will prepare graduate students for their target career by providing a customized education in which faculty and students work collaboratively to achieve a competency-based engineering education. To achieve this aim, we are piloting, testing, and validating a Personalized Learning Model (PLM) for STEM Graduate Education in which personalized learning is implemented both within the classroom and in professional development activities. The five components of the PLM include (1) establishing *Instructional Goals* for each student through learner profiles, strength finders, and individual development plans, (2) defining the *Task Environment* through one credit, modular classes that provide topic flexibility and content customization, and "Professional Development Streams" - a set of co-curricular activities organized around industry, academia and entrepreneurship, and (3) *Scaffolding the Instruction* to provide pedagogy that leads to independence and mastery in the student's area of focus. (4) *Assessment of Performance and Learning* tracks progression towards the student's instructional goals, followed by guided (5) *Reflection and Evaluation.* This talk will introduce the innovation, the evaluation and assessment approach, and describe progress to date. In the first model component – Instructional Goals – we have guided students in finding their strengths, establishing individual development plans, and effective goal setting; we will describe observations as a function of year in the program. In the second model component – task environment – we will describe the development and content of a modernized "body of knowledge" that incorporates input from disciplinary experts in industry, entrepreneurship, national labs, and academia. The qualitative and quantitative methods for formative and summative assessment will be presented. This work is supported by a grant from the National Science Foundation (DGE-IGE 2325599)

A Personalized Learning Model (PLM) for STEM Graduate Education

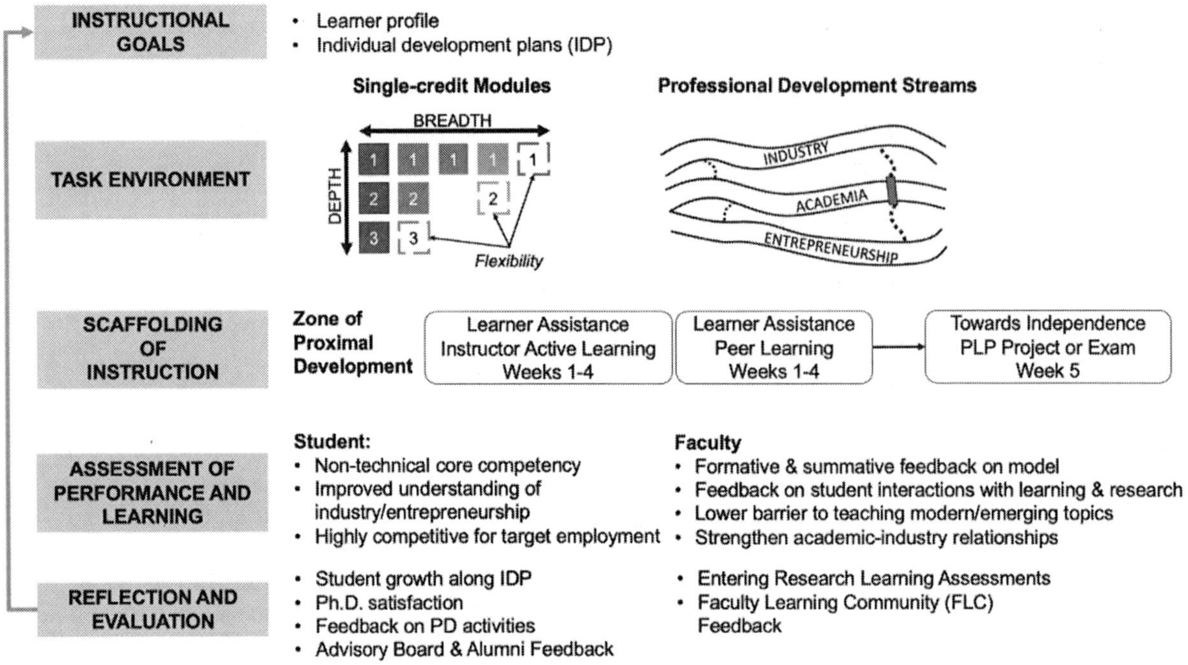

197f Checking the Bridge: Analysis of Grading and Structural Changes within Bridge Courses on Student Access to the Chemical Engineering MS

Christopher V. H-H. Chen, Gregory Chingas

The US Bureau of Labor Statistics (2023) has projected an 8% year-on-year growth in demand for chemical engineers–a much faster rate of growth than the average for all occupations (3%). This is in part due to how chemical engineers fit into addressing many of the societal goals the US prioritizes, from decarbonizing energy and developing greener food and water sources, to medical and health innovation and sustainable manufacturing. However, the rate of growth in the degrees awarded over the past 10 years (2012 to 2021) is not enough to keep with this demand, averaging 4.2% and 1.5% year-on-year growth for BS and MS degree completions, respectively in the US (Data USA, 2024). To address this need, our department has developed a program to allow non-chemical engineering undergraduates to access the chemical engineering MS by taking two "bridge" courses in their first term. This bridge course acts as a condensed post-baccalaureate overview of the undergraduate curriculum designed to prepare these students to access the graduate core chemical engineering courses by covering six topics in chemical engineering: chemical engineering math; material and energy balances, and controls; transport phenomena; thermodynamics; separations; and kinetics and reactor design.

Since Fall 2013, we have offered this bridge pathway to the MS, which has evolved over the years to better prepare our students in the program and to reflect changes that we have made to our undergraduate curriculum. We have also changed the grading scheme dramatically in 2022–moving from a traditional homework and exam-based assessment to a competency-based scheme–to better align the grading measures with the intents of the course: to prepare students to engage in graduate-level chemical engineering coursework by building their competency in undergraduate concepts. In the most current iteration, students who demonstrate competency in all course modules, by completing practice sets with over 95% accuracy, earn an A, with unlimited attempts to do so by the end of the course.

In this presentation, we share what we learned from our evaluation of these bridge courses, and check to see whether we are meeting the goal of facilitating competency in undergraduate chemical engineering skills and providing access to our graduate curriculum. We conducted a retrospective analysis on student grades in our MS program, and compared the performance of students in our graduate core courses who have and have not participated in these bridge courses from Fall 2018 through Spring 2024. We found mixed results in this analysis, where regular and bridge MS students performed statistically similarly in graduate Transport and Thermodynamics, but their performances were statistically different in graduate Math and Kinetics core courses (with regular students outperforming bridge students by 0.3-0.6 standard deviations). This statistical difference does not appear every year, however, and changes in the grading scheme (from traditional to competency-based grading) did not have a negative effect student grades in core courses. We also have compared the grade performance of students in the bridge courses to their performance in future graduate core courses and do not find that the bridge course grades are a strong predictor of future grades in the program.

The results of our analysis are shared to help show other departments interested in broadening access to their graduate programs how they may be able to do so, and to learn from our experiences running these bridging courses for over a decade. By checking our bridge, we continue to improve access to our own programs; adjust how we advise and support MS students who enter from outside of chemical engineering; and adapt how students experience and use these bridge courses to succeed in achieving a chemical engineering MS, allowing them to take advantage of the increasing job opportunities for graduates in our field that is expected over the next decade.

197e Improving the Validity of an Instrument to Measure Mental Health Help-Seeking Beliefs for Diverse Institutional Contexts

Sarah Wilson, Joseph H. Hammer, Gabriella Pitcher

Mental health differentially impacts undergraduate engineering students who are historically underserved in engineering. Further, engineering students who self-reported symptoms of mental health distress are significantly less likely to have sought professional help for their mental health. Increasing professional help-seeking behavior can improve the prognosis of mental health problems in emerging adults (i.e., 18-25 years of age) and reduce the potential for progression to more chronic or severe disorders. Therefore, this work aims to use stakeholder-driven research to identify mental health interventions across a College of Engineering by answering the following research questions: 1) What are the College specific barriers and facilitators to a culture of mental health? and 2) What do College stakeholders identify as key priority areas for transforming the culture of the College to be more supportive of mental health? Through Group Level Assessment (GLA) focus groups, participants identify and prioritize areas for improvement within a program and work together as the core stakeholders to identify next steps. This participatory action research model allows for the development of interventions that reflect the needs of the community while taking into account the diversity of the population. Additionally, inclusion of stakeholders in the development of interventions can lead to increased stakeholder buy-in, allowing for the development of more sustainable interventions. Through this talk, we will highlight the results of focus groups with approximately 40 engineering students. Outcomes of the focus groups include student identified interventions that aim to accelerate transformation of the College cultures towards being more supportive of student mental health.

59a Students' Perceptions on Sustainability in a Chemical Engineering Program

Pavan K. Inguva, Deesha Chadha, Umang V. Shah

Chemical engineers are well-placed to significantly contribute to many on-going and novel sustainability endeavors in society, leveraging the discipline's domain knowledge and skillsets. To that end, the undergraduate chemical engineering degree program needs to provide suitable learning opportunities to help students to acquire sufficient technical competencies, soft skills, and a sustainability mindset. The effective integration of sustainability into the chemical engineering curriculum raises multiple open questions that we as educators need to confront going forward. As part of an on-going curriculum review and transformation process at Imperial College London, a large-scale student survey (n=91) of all undergraduates (Year 1 – 4) was conducted to gauge students' perceptions on current department and program practices and to solicit ideas for future transformations. In this presentation, we discuss both qualitative and quantitative results from the survey which help to shed light on students' perspectives on sustainability and their educational experience. We also share some preliminary thoughts on areas for transformation and future work that are likely to be important in integrating sustainability into chemical engineering education.

59b Integrating Sustainability into Chemical Engineering Education through Project-Based Learning

Betul Bilgin

In response to the growing need for sustainable development and environmental stewardship in the chemical engineering sector, our CHE 201 Introduction to Thermodynamics course has implemented an innovative project-based learning (PBL) strategy aimed at embedding sustainability concepts into the curriculum. This initiative, structured around a comprehensive project where students work in groups, tasks learners with designing a sustainable product or process with significant market potential. The project encompasses several key phases: product selection, current state analysis, sustainable design brainstorming, design evaluation, modeling and prototyping, optional testing, and presentation.By granting students the autonomy to select products ranging from consumer goods to energy storage materials, the project emphasizes the importance of sustainability across various domains. Through a rigorous process that includes life cycle analysis, economic evaluation, and technical assessment, students engage deeply with the complexities of sustainable design. The use of SolidWorks for modeling and UIC MakerSpace technologies for prototyping further enriches the learning experience by providing hands-on exposure to state-of-the-art engineering tools.This educational approach not only enhances students' understanding of thermodynamics and its applications in real-world scenarios but also fosters critical thinking, creativity, and collaboration. The project culminates in a digital presentation using Adobe Express, facilitating a modern, accessible, and engaging method of showcasing their work. This talk aims to share insights and outcomes from integrating sustainability into chemical engineering education, contributing to the development of environmentally conscious future engineers.

59c An Integrated Approach to Enhance Global Awareness and Project Sustainability of Humanitarian-Based Projects

Randy Lewis, Noelani Mason, Terri Bateman

International educational opportunities for engineering students that promote global awareness include, but are not limited to, internships, study abroad programs, and humanitarian programs such as Engineers Without Borders (EWB). Since 2007, a two-semester Global Engineering Outreach (GEO) course in the College of Engineering at ____________ University has enabled engineering and technology students from multiple disciplines to design and implement humanitarian-based engineering projects. Projects have been implemented in Tonga, Ghana, Ecuador, Bolivia, and Peru. Current projects focus on prosthetics, cookstoves, and quinoa.

During the two semesters, teams composed of five to six students regularly interact with global community members during the design process. The teams are composed of at least three engineering and technology disciplines and students receive technical credit for course participation. Interaction is in Spanish since nearly half of the students speak and write Spanish fluently and have lived abroad in a Spanish-speaking country. This communication is important to establish relationships and to obtain feedback.

Recently, the global project experience was expanded to include other disciplines such as nursing, sociology, media arts, physics, food science, plant and wildlife sciences, and business. Currently, students and faculty from 11 majors and 7 colleges are participating. The vision for integrating students and faculty across campus is to break down silos and enhance project sustainability by 1) expanding inspired and integrated learning through work on global projects that are integrated across campus, 2) utilizing student strengths associated with living abroad and having foreign language abilities, 3) increasing connections to Latin America that benefit communities with humanitarian projects, and 4) integrating existing Study Abroad experiences.

In addition to enhancing project sustainability through integrating projects across many disciplines, project sustainability is enhanced by including three trips throughout the year. In August, a subset of faculty and students (10-20 individuals) travel to the communities in Latin America to follow up on projects that were implemented the previous May, implement any project updates, and formalize plans for the upcoming two-semester course. In the following January, a subset of faculty and students (10-20 individuals) travel to the communities to test prototypes, obtain feedback, and prepare for the large Study Abroad experience that will occur in May. In May, at the end of the two-semester course, all students in the course travel on an implementation trip as part of a Study Abroad program. Additionally, other Study Abroad programs (such as nursing or sociology) are arranged to overlap with the engineering Study Abroad program since these other disciplines have been working on the projects together.

Significant efforts during project development occur so that students can 1) receive training on social impact assessment and communication, 2) review past reports and videos to increase understanding of the culture, 3) interact on a regular basis with the community throughout the design process, 4) implement social constraints along with technical constraints in the design process, and 5) document all social interactions with the community.

This work will present an overview of the course structure, the integrated approach with other disciplines across campus, and program assessment and feedback. Additionally, global awareness and project sustainability strategies will be presented. Specific examples will be related to the prosthetic projects currently being implemented in Ecuador and the quinoa projects currently being implemented in Bolivia. An overview of how multiple Study Abroad programs are integrated together will also be presented.

59d Promoting Resource Management in Teaching Laboratories: A Case Study for Effective Implementation of Sustainable Practices

Umang V. Shah, Pavan Inguva, Mohamed Maher

Sustainability is on the agenda for organisations with growing public demand to address carbon emissions. Over the last three decades, the UK has reduced its emissions by 44% and is the first G7 to legislate the target of achieving net-zero carbon emissions by 2050 [1]. A recent report by The Royal Society of Chemistry [2] states that laboratories are responsible for 60 – 65% of electricity consumption within research institutes. In contrast, some research laboratories produce large amounts of waste, which accounts for 1.8% of the world's plastic production. Hence, teaching laboratories are energy and resource-intensive by nature, and one must take a holistic approach to alleviate the environmental impact while maintaining the teaching activities' objectives.

We present a case study of a teaching laboratory accommodating over 40 experimental and characterisation facilities and auxiliary equipment serving three core undergraduate modules. The equipment operates over 600 hours annually, leading to high energy, water and solvent consumption. We adopted a three-step innovative approach to reduce the environmental footprint of the teaching laboratory by decreasing the energy, water and solvent consumption.

First, we utilised a hybrid Photovoltaic/Fuel cell microgrid system to generate 16 kW and a closed-loop heating/cooling system that serves the entire laboratory. Secondly, a pilot-scale ion exchange facility recycles wastewater and produces demineralised water. Finally, an experimental ecosystem is employed to recirculate the solvents and significantly reduce solvent purchases. The ecosystem will substantially aid in reducing fresh solvent usage from non-sustainable sources while achieving the learning outcome of practical education in an authentic setting.

The results of incorporating the three vital sustainable practices into teaching laboratory operations show the potential for significant reductions in energy, water, and solvent consumption and enhancing the learning outcomes.

References

[1] Department for Energy Security and Net Zero (2022) Net zero strategy: Build back greener, GOV.UK. GOV.UK. Available at: https://www.gov.uk/government/publications/net-zero-strategy (Accessed: April 2, 2023).

[2] 'Sustainable Laboratories - Practices to Reduce the Environmental Effects of Research.' n.d. Royal Society of Chemistry. Accessed April 2, 2023. https://www.rsc.org/new-perspectives/sustainability/sustainable-laboratories/.

59e Sustainable Unit Operations Laboratory: Project-Based Learning Incorporating Life Cycle Analysis

Hariharan Sivanandam, Chijuan Hu, Ariel Chan

Carbon footprint evaluation is one of the standard approaches in estimating the emission of an enterprise or an institution. Most of the life cycle analysis is done for the whole building, and each activity's contribution is recorded and analyzed [1]. They considered all greenhouse gases and expressed all emissions in terms of CO_2 equivalent. These emissions are calculated based on the amount of activity and CO_2 emission intensity. The contribution of individual research activity or undergraduate laboratory teaching is not analyzed separately. The contribution of CO_2 emission from these lab activities is essential to provide best practices or action plans to reduce CO_2 emissions. However, this area has limited research to provide easy and robust approaches to analyzing carbon footprint and drive research and experimentation design. Note that research and educational activity in higher education differs from industrial settings in that the process usually operates at the most efficient point to maximize profits. In higher education, teaching/research activities aim to understand specific processes or fulfill the identified learning outcomes, so operation conditions are often varied and can be less economical.

Teaching laboratories are essential for higher education, providing students with hands-on experience and practical application of theoretical knowledge. However, these laboratories have a significant environmental cost, contributing substantially to energy and gas emissions, water consumption, and waste production. Recognizing this, efforts to minimize the environmental impact of laboratory operations have gained momentum. The Unit Operations Laboratory, in particular, stands out as a pivotal setting for students to not only grasp the theoretical aspects of their field but also to actively engage in sustainability practices. This laboratory, commonly utilized in science and engineering disciplines, allows students to understand the practical implications of their work on the environment and provides a unique opportunity to integrate Life Cycle Analysis. LCA is a systematic approach to evaluating the environmental impacts of a product, process, or service throughout its entire life cycle, from extraction of raw materials to end-of-life disposal. It assesses resource use, emissions, and energy consumption to identify areas for improvement and inform decision-making.

This study presents the case study of bioethanol production via yeast fermentation, incorporating the LCA into undergraduate laboratory problem-based projects. The undergraduate and master's students (taken toward their degree as the lab courses or thesis study) investigated CO2 emission in the existing operations, proposed alternative production pathways, and the subsequent distillation operation conditions. The results contributed to the experimental design of future projects for the undergraduate lab courses. The LCA analysis is conducted based on Scope 1 and Scope 2, following ISO 14040 and 14044 standards, using Defra GHG emission factors for 2022.

With a focus on the environmental impact of laboratory equipment operation, this paper/presentation aligns with global sustainability initiatives, clean energy efforts, and the commitment to achieving net-zero emissions by 2050. The findings show that using different substract can significantly reduce CO_2 emission (ie. refined sugar with yeast for bioethanol production reduces the carbon footprint by half compared to corn syrup feedstock). For the subsequent purification process, simply varying the steam heating pressure for the reboiler can reduce the CO_2 emission by 20-30%, depending on the reflux or flash operating conditions. Since there is variability in laboratory practices across different regions and institutions, international collaboration is necessary to test the robustness and scalability of common unit operations laboratory projects. For future work, collaborative efforts can provide diverse perspectives, validate findings, and contribute to the development of standardized practices in laboratory education. Two institutions, one in Canada and one in the US, worked together to test the LCA approach on four distillation columns, a 20 ft plate and packed distillation column in Canada and 35ft plate and packed distillation columns in the US, and 2 bioreactors of 80 L in another institution and 80 L in another one. Ultimately, this research work contributes to the broader goal of embedding sustainability principles within the educational framework, fostering a generation of professionals equipped with the knowledge and skills necessary to address environmental challenges.

59f Integrating Sustainability across the Chemical Engineering Curriculum

Courtney Pfluger, Christopher V. H-H. Chen

The United Nations 2030 Agenda for Sustainable Development represents a global commitment to addressing the world's most pressing economic, social, and environmental challenges with 17 goals. Specifically, goal 12 looks to address sustainable consumption and production patterns that have many ties to the chemical engineering field. Therefore, teaching chemical engineers to consider and evaluate the impacts of their designs on the environment, individuals, and society is imperative to achieve this Sustainable Development Goal. This need has become so important that ABET has even began to include these considerations in Student Outcome 2, which currently includes the "...consideration of public health, safety, and welfare, as well as global, cultural, social, environmental, and economic factors." In this presentation we will discuss how chemical engineering educators from Columbia and Northeastern Universities, two large, private, R1 universities, have introduced and engaged students in considering sustainability. We combined elements from Engineering for One Planet Framework, the Professional Social Responsibility Development Model, and Global Competencies frameworks into three compact Sustainability-in-Action Elements (SAEs):

- *Sustainability Awareness*: Students will be able to describe and appreciate different cultures, values, and perspectives; the sources of pollution, emissions, and climate change; and the need for more sustainable practices.
- *Evaluating Impact*: Students will be able to determine the impacts on society and the environment of an engineering action.
- *Taking Action*: Students will be able to apply their sustainability awareness and ability to evaluate impact to design an engineering solution that creates positive environmental and social value.

We collected quantitative and qualitative data from the sustainability-informed student learning objectives in three courses–a Material & Energy Balances course for sophomores, a senior capstone design course, and a senior elective on green chemical engineering. We will demonstrate how students learned and applied sustainability values through their engineering design projects in these courses. Data of the self-assessed knowledge from these learning outcome surveys will inform how well the courses addressed sustainability topics, the outcomes for ABET Student Outcome 2, and how the students felt about the applications of these topics in the courses. The following questions will be addressed: (1) How can chemical engineering educators better address sustainability topics in chemical engineering courses? (2) How could chemical engineering educators implement sustainability topics throughout the chemical engineering curriculum to improve students understanding of the need and ability to apply these topics?

59g Expanding Sustainable Engineering Curricula Via Existing Frameworks to Meet the Needs of Students

Jennifer Watt, Debolina Banerjee, Adrienne Cachelin, Anthony Butterfield, David R. Wagner

Sustainable Engineering Education Development (SEED) is a new program at the university that integrates concepts of sustainability into engineering curricula. We work with the university's Sustainability Office to develop assessment strategies for sustainability education campus-wide. The Engineering for One Planet (EOP) Initiative provides the framework and learning outcomes for the program's goal to expand the instruction of sustainability concepts to every engineering discipline.

The EOP Initiative and Framework have already provided an excellent foundation we intend to use for assessment. The EOP Resources webpage contains 93 core and advanced learning outcomes to help move knowledge forward. A matrix has been constructed to track the implementation of these outcomes in completed and upcoming courses. This matrix will also aid ABET accreditation efforts and incorporate global considerations via the United Nations Sustainable Development Goals. Making these connections will allow students to see the global significance of their education and decisions. In addition to these sources, the National Academies Press has published a report entitled "New Directions for Chemical Engineering," in which they detail recommendations for research, improved educational tools, training, and leadership. Our educational plan includes at least five recommendations that focus on advancing training and leadership by combining each team member's experiences in education and research.

To complement the quantitative evaluation, the Sustainability Office and Department of Chemical Engineering will qualitatively assess the impact of EOP instruction on students in representative EOP-infused courses; resources already exist for these assessments. These assessments utilize focus groups of students and a group of faculty. Focus group questions for students and faculty have been developed and focus groups are nearly complete. Student focus groups will center on what they have learned and how they use that information in their degree as well as outside of coursework. Faculty focus groups will concentrate on barriers to instruction in attempts to identify apprehension of adding content to existing course. Implementation of concepts of sustainability in new courses will also be addressed; a new elective will be taught in the summer of 2024 entitled "Introduction to Sustainable Engineering."

59h Is It Time to Survey How We Teach Sustainability in Chemical Engineering?

Joaquin Rodriguez Alonso, David V. P. Sanchez

AIChE has successfully developed the strategy of surveying the core courses of the chemical engineering curriculum every ten years [1-8]. These surveys have explored how faculty around the country update the teaching of Introduction to Chemical Engineering, Mass and Energy Balances, Thermodynamics, Transport Phenomena, Reaction Engineering, Process Control, Process Design, and Unit Operations Lab courses, considered to be the backbone structure of the chemical engineering career. The inclusion of Safety and Professional Electives has come to round up those other topics considered someway essential for the professional formation [9-10]. The analysis of those surveys has proved very insightful for AIChE to keep track of the state of academia in chemical engineering, and for faculty and other interested stakeholders to monitor the relevance of the content and strategies in the formation of the next generations of chemical engineers. The emergence of sustainability mainly in the shape of professional electives followed a much encouraged and supported initiative starting with the promotion of the topic from UN conferences on the environment [11-23]. It was further reinforced with the program of the Decade for Sustainability Education [24]. However, the integration of sustainability in the chemical engineering curriculum has adopted much dispersed strategies and some relatively slow pace. Several attempts have been tracking the progress of these initiatives [25-26]. Led by rapid changes in research and industrial developments, the "topic" has already become and essential skill, not to say a core formation, for new professionals in the field and most certainly for the next generations of chemical engineers. This fact calls the attention about the central role that sustainability should be playing in an updated chemical engineering curriculum and the strategies to turn it into effective teaching methods. A preliminary web survey of top chemical engineering departments in the U.S. and some scattered data from similar European institutions reported here reveals a tremendous effort at offering content on sustainability at higher education level (undergraduate, graduate) in the form of elective courses, extending into concentration or certificate programs, and even master and doctoral programs. It seems, however, that the rich diversity of these initiatives could provide for a more integrative approach consolidating a global strategy to core teaching in chemical engineering. It will also help in guiding more institutions to adopt an efficient inclusion of sustainability. In this regard, to add an example of what is being done and what needs to be done, our institution has developed a strategy based on the establishment of an interdisciplinary center for sustainable development, increasingly involving engineering disciplines (starting with a nucleus of civil and environmental engineering) and more recently adding the sciences, humanities, business, and medicine. The center supports research and teaching, offers certificate programs for students and professionals, and serves as an encounter plaza for organizations, government and private institutions, and academia to discuss, rise and examine proposals on a broad range of sustainability efforts. Schools and departments add their own initiatives (mainly as elective courses) fitting particular interests of their academic programs. The chemical engineering department (a main contributor to the origins of the center referred to above) has recently started to develop its own strategy to incorporate sustainability into the curriculum [27] with an elective course on "green chemical engineering" for undergraduates and an elective course on "process intensification" for graduate students. In combination with an introductory course on sustainability, by chemical engineering faculty, and the leadership of a doctoral program in circular economy, provides a simple "linear" path (from first-year student to a Ph.D. degree) in teaching sustainability. However, it would benefit extensively from a more consolidated approach and insightful advice from faculty considering the need to include sustainability as a "core" formation for chemical engineers, as a timely survey could support.

[1] Ford, L. P., Brennan, J., Cole, J., Dahm, K. D., Jamieson, M. V., Landherr, L. J., Silverstein, D. L., Vaughen, B. K., Vigeant, M. A., Thiel, S. W. (2020, June), *How We Teach: Chemical Engineering in the First Year* Paper presented at 2020 ASEE Virtual Annual Conference Content Access, Virtual On line . 10.18260/1-2—34738, https://peer.asee.org/34738

[2] Ford, L., Brennan, J., Dahm, K., Silverstein, D., Landherr, L., West, C., Cole, J., Thiel, S., Vaughen, B., Jamieson, M. (2022, August), *How We Teach: Material and Energy Balances* Paper presented at 2022 ASEE Annual Conference & Exposition, Minneapolis, MN. 10.18260/1-2—41893, https://strategy.asee.org/41893

[3] Vigeant, M. A., Cole, J., Dahm, K. D., Ford, L. P., Landherr, L. J., Silverstein, D. L., West, C. W. (2019, June), *How We Teach: Thermodynamics* Paper presented at 2019 ASEE Annual Conference & Exposition , Tampa, Florida. 10.18260/1-2—32903, https://peer.asee.org/32903

[4] Lepek, D., Vigeant, M. A., Silverstein, D. L., Keith, J. M. (2015, June), *How We Teach: Transport Phenomena and Related Courses* Paper presented at 2015 ASEE Annual Conference & Exposition, Seattle, Washington. 10.18260/p.24201, https://peer.asee.org/24201

[5] Ford, L. P., Brennan, J., Silverstein, D. L., Landherr, L. J., West, C. W., Thiel, S. W., Dahm, K. D., Cole, J., Jamieson, M. V. (2021, July), *How We Teach: Kinetics and Reactor Design* Paper presented at 2021 ASEE Virtual Annual Conference Content Access, Virtual Conference. 10.18260/1-2—37261, https://peer.asee.org/37261

[6] Silverstein, D. L., Vigeant, M. A., Staehle, M. (2016, June), *How We Teach Process Control: 2015 Survey Results* Paper presented at 2016 ASEE Annual Conference & Exposition, New Orleans, Louisiana. 10.18260/p.25495, https://peer.asee.org/25495

[7] Ford, L. P., Cole, J., Dahm, K. D., Vaughen, B. K., Jamieson, M. V., Landherr, L., Silverstein,, D. L., Vogel, T. J., West, C. W., Thiel, S. W. (2023, June), *How We Teach: Capstone Design* Paper presented at 2023 ASEE Annual Conference & Exposition, Baltimore , Maryland. 10.18260/1-2—43394, https://sftp.asee.org/43394

[8] Vigeant, M. A., Silverstein, D. L., Dahm, K. D., Ford, L. P., Cole, J., Landherr, L. J. (2018, June), *How We teach: Unit Operations Laboratory* Paper presented at 2018 ASEE Annual Conference & Exposition , Salt Lake City, Utah. 10.18260/1-2—30587, https://peer.asee.org/30587

[9] Ford, L. P., Carter, T., Brennan, J., Dahm, K., Barr, C., Flodman, H., Landherr, L., Silverstein, D., Thiel, S., Vaughen, B., Vogel, T. (2023, November), *Process Safety across the Curriculum* Paper presented at 2023 AIChE Annual Meeting, , Orlando, Florida (129a) https://www.aiche.org/conferences/aiche-annual-meeting/2023/proceeding/paper/129a-process-safety-across-curriculum

[10] Ford, L. P., & Cole, J., & Dahm, K., Brennan, J., Silverstein, D., and Thiel, S. (2023, November), *How We Teach: Chemical Engineering Electives* Paper presented at 2023 AIChE Annual Meeting, , Orlando, Florida (129b) https://www.aiche.org/conferences/aiche-annual-meeting/2023/proceeding/paper/129b-how-we-teach-chemical-engineering-electives

[11] United Nations Conference on the Human Environment, 5-16 June 1972, Stockholm. https://www.un.org/en/conferences/environment/stockholm1972

[12] Report of the World Commission on Environment and Development: Our Common Future.

1987. https://sustainabledevelopment.un.org/content/documents/5987our-common-future.pdf

[13] UN Conference on Environment and Development, Rio de Janeiro, 5-14 June 1992 https://www.un.org/en/conferences/environment/rio1992

[14] Special Session of the UN General Assembly, New York, 23-27 June 1997 https://www.un.org/en/conferences/environment/newyork1997

[15] Millennium Summit, New York, 6-8 September 2000 https://www.un.org/en/conferences/environment/newyork2000

[16] World Summit on Sustainable Development, Johannesburg, 26 August – 4 September, 2002 https://www.un.org/en/conferences/environment/johannesburg2002

[17] World Summit, New York, 14-16 September, 2005 https://www.un.org/en/conferences/environment/newyork2005

[18] High-level Meeting on the Millennium Development Goals, New York, 22-25 September, 2008 https://www.un.org/en/conferences/environment/newyork2008

[19] Millenium Development Goals Summit, New York, 20-22 September, 2010 https://www.un.org/en/conferences/environment/newyork2010

[20] UN Conference on Sustainable Development, Rio de Janeiro, 20-22 June, 2012 https://www.un.org/en/conferences/environment/rio2012

[21] Special Event towards Achieving the Millennium Development Goals, New York, 23-25 September, 2013 https://www.un.org/en/conferences/environment/newyork2013

[22] UN Summit on Sustainable Development, New York, 25-27 September, 2015 https://www.un.org/en/conferences/environment/newyork2015

[23] Stockholm+50 Conference, Stockholm, 2-3 June, 2022 https://www.stockholm50.global/

[24] United Nations Educational, Scientific and Cultural Organization. Shaping the Future We Want. UN Decade of Education for Sustainable Development (2005-2014). Final Report. UNESCO 2014. https://sdgs.un.org/sites/default/files/publications/1682Shaping%20the%20future%20we%20want.pdf

[25] Glassey, J.; Haile, S. *Sustainability in chemical engineering curriculum.* International Journal of Sustainability in Higher Education 13(4): 354-364, 2012

[26] Sheehan, M., Desha, C., Schneider, P., and Turer, P. *Embedding sustainability into chemical engineering education: content development and competency mapping*. Chemeca, 2012 https://researchonline.jcu.edu.au/26335/1/Chemeca_2012_Paper__Sheehan_et_al.pdf

[27] Reference to be provided after approval to keep the anonymity of the authors during the process

253a Improving Cybersecurity Awareness By Student Group Interactions on a Process Control Laboratory Distributed Control System

Peyton Richmond, Helen Lou, Tianxing Cai

This presentation describes our experience developing a new activity to demonstrate a process control engineer's role and raise awareness of the risks posed by industrial cybersecurity breaches in chemical manufacturing facilities. Lamar University's Process Control Laboratory has seating for four groups of students to operate distributed control system (DCS) automated experiments simultaneously. Typically, three student groups each operate an experiment through the common DCS which provides them with the typical functionality of chemical manufacturing facilities. This year a process control engineer student group was added. This group was assigned to diagram the DCS block configurations used to control the three operating group's experiments. After diagramming the controls, the fourth group was then directed to surreptitiously interfere with the DCS controls of the three operating experiments. Thus, the three operating groups must constantly assess whether the controls to their experiment's are being hacked.

The process control engineer group simulates cybersecurity issues for the three operating groups by temporarily modifying online parameters of the DCS function blocks that control each experiment. They choose from the following actions: (1) manipulate the operating group's valves or pumps using a digital or analog output, (2) spoof a reading for a digital or analog input, or (3) make changes to operator accessible items. If the targeted operating group correctly detects the unwanted interference and calls out the process control group, control will be returned immediately. However, if the operating group does not detect the changes or cannot determine what the control group is doing to interfere with their experiment then they must put up with the activity for up to five minutes. We found that these activities were extremely effective at capturing our student's interest and seemed to be very enjoyable.

253c Safe Take-Home Experiments for Chemical Engineering Education

Kang Zhou, Wean Sin Cheow

In this talk, we will present a recent pilot case study we led in the Singapore Institute of Technology. The study aimed to deploy experiential learning in a Year-1 Python Programming course with a class size of 114 students. Approximately, 40 low-cost, lunch box-sized devices were assembled for students to appreciate the concept of automation. Each device was a plastic box, containing a micro-controller, a CO_2 sensor, a fan, and a bowl (cost of each device: ~40 USD). In a group project (size of most groups: 3), students took the device home and turned them into CO_2 incubators. They were instructed to prepare a solution containing yeast and sugar in the bowl to produce CO_2. The micro-controller has been programmed for them, so that students could send simple commands from their laptop to the micro-controller via Serial communication for requesting CO_2 data and controlling the ventilation fan. Most groups have met the technical specifications (maintaining CO_2 in the range of 4,000 ± 200 ppm for at least one hour). In the project, students practiced the basic programming techniques they learnt in the theory part of the course, especially recursion, conditional statement, various variable types, data management using DataFrame, and data visualization. They also experienced how a closed-loop control system works and found the "tricks" needed to manage the deviation from the setpoint. Many of them have used proportional control after they realized that the On-Off control could not meet the objective. They have also witnessed a steady-state error and could not address it without manually offsetting the setpoint. This experience should help them understand and appreciate the PID control theory in their Year 3 course. The project also had an open-ended component where they were tasked to design interesting experiments using the device. The official student feedback report is not available when this Abstract was submitted, but some students actively approached the lecturer to express their positive feedback on the project.

We will also briefly discuss how the modular device can be expanded to conduct more experiments outside standard labs for Chemical Engineering education. It will rely on biocatalysis to carry out experiments under mild conditions and utilize low-cost sensors (e.g., light, temperature, weight, pH, camera) and actuators (e.g., LEDs, peristatic pump, heating element, agitator). Such activities could complement the existing standard lab sessions and allow students to have more hands-on activities with less time and resources constraints.

253d Exploring the Power of Desk-Scale Experiments for Enhanced Understanding of Flow Characteristic Curves

Fernando Mérida Figueróa, Sindia M. Rivera-Jiménez

This work presents an innovative approach to studying flow characteristic curves (CUR) through desk-scale experiments, aiming to enrich student learning objectives and diversify traditional chemical engineering laboratory structures. The main goal is to assess the effectiveness of desk-scale experimental kits in enhancing large-scale experiments in unit operations laboratories, particularly for remote learning modalities. The study seeks to address this question: "How learning objectives and pedagogical approaches can be improved to facilitate students' comprehension of key fluid mechanics concepts via CUR desk-scale experiments?

The CUR desk-scale module comprises aquarium pumps, feed tanks, valves, 3D-printed pressure taps, and other elements connected by flexible tubing. A differential pressure sensor connected to an Arduino microprocessor quantifies pressure drop across pumps and valves, which is used to construct characteristic curves for various types of pumps, valves, and pump arrays. The primary learning objectives focus on reinforcing fluid mechanics topics not fully covered in lecture courses, via experimentation within the Unit Operations 1 laboratory course at the University of Florida.

Student-collected experimental results demonstrate that desk-scale CUR experiments effectively replicate larger-scale pump and valve behaviors, aiding in understanding challenging concepts such as predictive system curves, net positive suction head, and the additive nature of pressure head and flow rate in pump arrays. Notably, this enhanced understanding was achieved with only minor adjustments in the configuration of the small experimental setup. These findings highlight the pedagogical value of desk-scale experiments, emphasizing their simplicity and accessibility in teaching chemical engineering principles.

This study demonstrates that CUR desk-scale experiments play a pivotal role in chemical engineering laboratories, offering invaluable learning opportunities. Their integration into remote learning modalities should be guided by positive student feedback, underscoring their significance in enhancing engagement and comprehension of fluid mechanics topics. The use of interactive elements along with clear instructions mitigate technical challenges has been identified by students and must be addressed. In-person pilot-scale experiments further enrich learning by offering access to industry-standard equipment and hands-on guidance. This blend fosters collaboration, responsibility, and mastery of experimental procedures, thus contributing significantly to educational transformation in chemical engineering laboratories.

253e Low-Cost Hands-on Micro Spectroscopy Experimental Module for Chemical Engineering Education

Riley Fosbre, David B. Thiessen, Prashanta Dutta, Olusola Adesope, Bernard Van Wie

The focus of the project is to teach prospective engineers about microfluidics and microtechnology. We wish to do so interactively to visualize the topic being studied for students, providing a complete model of the phenomena to address common misconceptions. Microfluidics was the topic chosen to be addressed by this project due to its importance, complexity, and relative obscurity in the current paradigm of engineering education. Microfluidics is de ned as the fluid behavior associated with the processing of micro-scale quantities of fluids and is thus a vital topic for those in the field of biochemical engineering and analytic science. Microfluidics has radically impacted chemical and biological analyses, allowing for improved understanding of cellular biology, point-of-care medical diagnoses, and of course fluid mechanics, while allowing for the production of economic and disposable micro-devices due to the advancement in soft lithography techniques. However, despite the topic's relevance to many engineering disciplines, including microfluidics in a traditional engineering academic track has proven difficult, as it is multidisciplinary in nature, deriving concepts from physics, material science, chemistry, and biology.

There is currently a serious gap in engineering curricula involving microfluidics, so in the long term we intend to promote a new educational pedagogy to be adopted by the larger community to address this gap. In the short term, we attempt in this project to address this shortcoming through the development of low-cost desktop learning modules for classroom and laboratory usage. Educational materials will be provided with these modules, and the effectiveness of these new pedagogical implementations will be assessed through a careful examination of student comprehension of the material provided.

The first module under development to meet this end is a micro-scale glucose assay analyzer, which relies on microfluidic principles and a common biomedical topic, glucose, to cover several scientific topics at once: reaction kinetics, absorption spectroscopy, and Beer's Law.

Glucose is commonly measured as a blood chemistry because it is a key indicator for many medical conditions, such as diabetes. The concentration of glucose in blood is commonly measured via a multi-stage reaction in which glucose is first oxidized to produce peroxide, which is then detected by reacting with 4-amino antipyrine in the presence of a different enzyme, peroxidase, which in turn yields a final product, quinonimine, a red dye with a characteristic wavelength of light it absorbs. Under optimal circumstances, the concentration of glucose in the solution can be correlated to the amount of light allowed to pass through based on Beer's Law, as it will be proportional to the amount of peroxide produced, which is in turn proportional to the amount of quinonimine present in the sample after the reaction. While common in the biomedical sector, use of this reaction allows instructors to cover enzyme-mediated reaction

kinetics, stoichiometry, and basic spectroscopy, making it ideal for an educational setting.

The experimental set up is simple. Students will inject sample and reagent into respective legs of a Y-shaped chip held in the vertical position, with an on-off valve at the intersection of the Y. From there, the sample will be mixed through a serpentine micro-channel static mixer as flow is induced by gravity, resulting in shades of red depending on the concentration of glucose in the sample (faint pink for low concentration, deep red for high). From there, the sample will pass into a reservoir with two sides. One side is covered with a polarized film, intended to allow only a narrow bandwidth of green light to pass through. This side of the reservoir is backlit, allowing green light to pass through the sample, which is then absorbed by the red dye present in the sample. From there, the green light that manages to pass through to the other side of the reservoir will be measured using a smartphone camera, and with Beer's law, students can determine the concentration of glucose in the sample based on the measured intensity and a calibration curve. Students will be guided through the experimental proofs by using accompanying worksheets. Comprehension of related concepts will be tested where the reaction kinetics and scientific principles will be tied in with the course. This experiment will work well within biomedical and reaction engineering courses.

The Prototype for dissemination is still in the conceptual phase and is currently in development. Since prototyping moved passed the proof-of-concept version, different manufacturing methods are being considered for the optimal route for mass production, e.g., traditional 3D printing, resin printing, laser cutting, etc. At present the concept is to form the module as a three-piece design composed of acrylic for transparency, with the micro-channels machined into the center before being fused to front and back panels using a UV cured epoxy. The module would be gravity fed, with holes drilled through the front and back panels at the bottom of the module, just above the reservoir, to provide an outlet for air and for easy washing after experimentation. This proposed version of the design minimizes the number of components per kit in favor of a single pass, gravity fed, set-up with the only electronic components being the student's cellphone and a light source for back lighting. The hope is this will limit manufacturing and maintenance costs, so that the bulk of expenses can be received for experimental materials, such as the glucose and enzyme solutions.

253f From Following to Developing: A New Approach to Student-Developed Lab Manuals for Unit Operation Laboratory Courses

Khaled Elsaid, Arvin Arugay, Raid J. Hassiba, Marcin Kozusznik, Ghada Salama

Laboratory courses in general, and unit operation laboratory courses in particular are very crucial in the chemical engineering curriculum, as they present the best direct demonstration of many theoretical principles and phenomena. Accordingly, high attention has been given to the effective teaching and delivering methods related to these courses. At the heart of the laboratory courses is the lab manual, which provides the necessary background and instructions for students to successfully perform their lab experiments, hence obtaining the expected outcomes. However, lab manuals are usually developed by the course instructors in a similar fashion to lecture notes, while considering the lab-specific nature and students' knowledge level. In the current work, we have explored the potential of having the lab manual developed by students themselves, i.e., a student-developed lab manual, so students are developing rather than following the lab manual. We have asked a group of senior students to work on developing a lab manual for a newly introduced experiment, which is to be incorporated into the curriculum for the semester to follow. A lab setup for the demonstration of adsorption/desorption through humidification/dehumidification of air on silica gel has been chosen for this purpose. The students have been provided with the manual for the different components of the lab setup, and the main outlines for the experimental objectives. Then students were asked to develop the experiment manual to fully explore the process and the effect of different operating variables on the process output. The experiment manual developed by students has undergone several review rounds with the lab instructor and lab coordinators to provide constructive feedback to improve the manual quality and make sure that fundamental experiment objectives are clearly communicated. Furthermore, the procedure must be correct and in proper order with no confusion for the safe operation of the equipment. The observations have revealed that students have been able to generally develop a good preliminary experiment manual at an acceptable level. However, this manual has to undergo a rigorous and iterative review process with in-depth feedback and input from the course instructor. Meanwhile, it was found that the student's developed manual is generally simple and easy to understand as it uses terms and descriptions easy to absorb at the student's level of knowledge and understanding. The other main findings and observations are presented in this paper.

253g Practical Design of Experiments for the Next Generation of Semiconductor Process Engineers

Sumant Patankar

Production of the latest high-value semiconductor chips will rely on development of cutting-edge tools and processes. One of the primary goals of the recent CHIPS act (Chips and Science act, 2022) is to accelerate workforce development for the semiconductor industry, as it is expected to add hundreds of thousands of jobs. Chemical engineering is a profession geared towards the manufacturing industry. The ability to decode complex chemical processes, design equipment, and implement process controls puts chemical engineers at the forefront of safety, process development, and quality control in the semiconductor industry. However, core chemical engineering curricula do not expose students to the unit operations of semiconductor manufacturing such as implantation, lithography, etching, and film growth/deposition.

At the University of Florida, both at the graduate and undergraduate level, requisite lab modules provide hands-on exposure to semiconductor unit operations and design and development of stable semiconductor processes. At the undergraduate level, these modules are included in the senior-level . For the undergraduate curriculum, the objective is to provide exposure to semiconductor processes; as such, the students follow a standard lab manual with recommended experimental conditions. At the graduate level, master's students learn about semiconductor processing and design of experiments in a required lab course (Advanced Chemical and Bioprocessing Lab). This lab course focuses on diverse areas of chemical engineering such as semiconductor manufacturing. Identifying stable process conditions for semiconductor processing is emphasized through use of design of experiments (DOE) techniques.

The graduate-level photolithography lab module aims to provide practical exposure to real-world scenarios by letting students perform a process optimization exercise using DOE and statistical tools. In this paper, graduate students attempted to target the lithography process to a feature size of 10 micro-meters using two experimental design methods - 1. OFAT (One Factor At a Time) and 2. Factorial design.

In addition to photolithography, the lab course also includes modules on dry etching, wet etching, thermal oxide film growth, metal vapor deposition, and metal plasma sputtering.

253h Flavor Analysis of Beer in the Chemical Engineering Laboratory Course

Evelyn Schwartsman, Benjamin Davis, Jennifer Weiser

In 2023, the global craft beer market size increased by 5% (to $28.4 billion), now accounting for 24.6% of the $115 billion U.S. beer market [1]. With the brewing industry experiencing significant growth, it is important for chemical engineering students to gain practical experience in brewing techniques. This not only familiarizes them with process control and parameter sensitivity but also exposes them to complex bioreactor operations that would be encountered in various industrial sectors.

A new addition to an existing beer brewing project within a senior lab class of the chemical engineering curriculum will involve a specific focus on hops (a critical ingredient in brewing that impacts flavor of final products). The experiment allows students to explore this additional parameter into an inquiry-based approach, with decisions such as quantity and timing of the addition of hops left for the students. They would then analyze their final samples through High Performance Liquid Chromatography (HPLC) to quantify the presence of iso-alpha acids responsible for flavor profiles. Using this method, students can directly correlate the chemical compositions of the samples to known flavor profiles of beer products using prepared calibration curves and known standards.

This exercise is part of a broader learning objective for senior students to understand the brewing process, which includes steps like milling, malting, wort filtration, and fermentation. This incorporation of HPLC into a laboratory project focusing on the beer brewing process presents a consistent and reliable approach to enhance chemical engineering education, specifically in understanding the role that hops play in flavor profiling of brewing; by comparing their results to established standards, students will not only gain lab skills but also ensure product quality and consistency in the brewing industry. By allowing this hands-on approach to experimenting with flavor profiles, the students' understanding of the brewing process and the relevance of chemical engineering practices in the food industry underscores the importance of tasks like parameter optimization and industrial bioreactor control.

[1] National Beer Sales & Production Data. Brewer's Association (2022).

521a A New Approach to Professional Development of Students By Integrating Aiche Resources in Chemical Engineering Seminar Classes

Gabriela Nagy, Karissa Raderstorf, David Corti

Seminar classes are an integral part of chemical engineering education at Purdue University and they are taken as required one credit hour courses at sophomore, junior and senior level. Each level course helps students learn more about the chemical engineering profession and introduces them to areas of practice, career choices, professional ethics, safety and valuable skills as they prepare to enter the professional world.

While the traditional format of the seminar class, with lectures on relevant topics is preferred for the sophomore and junior level courses, a new approach to the senior level course was recently introduced to allow for tailored professional development for each student. The goal of the senior level seminar course is to help prepare students for the transition from being an undergraduate student to becoming a successful professional or graduate student. In the new format, the course is offered as individual study, with students taking ownership of their professional development by completing activities (in-person or online) and earning points towards a minimum/maximum threshold in four strategically chosen categories, to fulfil ABET requirements: (1) Career Readiness and Life-Long Learning, (2) Contemporary Issues in Chemical Engineering, (3) Engineering Ethics, and (4) Culture, Equity, Diversity and Inclusion.

With this new course format, the students have the flexibility to complete a variety of activities for credit, ranging from in-person events, seminars or workshops organized on- or off-campus, to watching relevant videos/recorded webinars, reading articles or books or meeting with mentors. The students are provided with list of resources in each category, and are encouraged to use the resources freely available through the AIChE Academy, including recorded conference sessions, Safety and Chemical Engineering Education (SAChE) and other online courses, articles, etc. Students are also encouraged to find other activities in line with their interests and complete them for credit after they are vetted by the instructor.

This presentation will describe in more detail the structure of the course, will present examples of activities completed for credit in each category and will review the feedback received from students on this new course approach.

The feedback received from students is highly positive; the new format allows for flexibility with the time when they complete activities, and the type of activity completed. The list of resources and the types of activities is reviewed regularly and expanded throughout the semester to provide more opportunities for professional development for enrolled students.

521b A Workshop to Help Students Build a Great Student Chapter Using an Entrepreneurial Mindset

Matthew Liberatore

Being involved with larger organizations can help us grow as students, professors, or working engineers. Professional engineering organizations, specifically AIChE here, offer opportunities to engage and lead during undergraduate study (student chapters) as well as throughout a person's career (local sections, divisions, forums, etc.). Here, several iterations of a goal-setting and leadership-development workshop focused on student chapters will be profiled. The overarching objective of the workshop is to lead participants through exercises to help build great student chapters. Specifically, AIChE students at a regional conference, STEM students and advisors on the author's home campus, and student chapter advisors at the AIChE Annual Conference participated in different activities centered upon an entrepreneurial mindset framework. This entrepreneurial mindset framework is summarized by the 3C's, namely Curiosity, Connections, and Creating value. By applying an entrepreneurial mindset, participants identify characteristics desired in their student chapter, officers, and members. Additional workshop prompts explore actions of student members and leaders and help participants develop an action plan for the coming semester and year. The logistics and delivery of the workshop will be detailed. In addition, surveying participants at both the beginning and end of the workshops found high engagement, i.e., significant evidence of brainstorming and goal setting. Workshop resources, such as slides and worksheets, will be provided, so other faculty or chapter advisors can deliver this workshop in other settings. This contribution will build upon a recent ASEE conference paper on this topic.

521c Exploring Technical and Ethical Decision-Making in Chemical Engineering through a Multi-Week Choose-Your-Own Path Simulation

Beth Rundlett, Jennifer Fiegel

Engineering decisions carry substantial societal implications, both beneficial and potentially detrimental. Therefore, it is crucial for engineers to be able to identify, comprehend, assess, and alleviate adverse impacts. To help equip students for their future professional roles, they must be trained to understand the technical, ethical, and societal ramifications of their engineering choices.

Toward this goal, we have previously developed an interactive, team-oriented activity that immerses engineering students in ethical decision-making based on a real-world engineering design failure that manifests after launch of a product. Two scenarios, an automobile part failure and a biomedical device failure, were developed with multiple potential pathways and unique outcomes. This choose-your-own-path activity requires students to make multiple decisions that have ramifications for themselves, their companies, and their future. However in the game, they also must contend with decisions made by hypothetical colleagues, management, and customers that they may or may not agree with. This work was aimed at early engineering students in their first or second year in college who have just begun to develop their technical skills, with the goal of helping undergraduate students see that engineering decisions are complex and can have larger consequences than originally anticipated. Results from this activity, which has been completed by over 1500 engineering students in the past five years, show a high level of engagement from the students and that students note the difficulty of decision making where there are many considerations.

In the current work, we present an expanded module based on the Flint water crisis for chemical engineering students, where they have to use their acquired chemical engineering skills to help them decide the right path forward. This case study is less well defined than the previous scenarios (more similar to the engineering problems they would face in their careers), requiring students to use their engineering judgment, ability to gather and collate data from disparate sources, and their mathematical skills to proceed. It also spans multiple weeks, rather than being a single 50-min experience. Students are required to complete activities such as calculating flow rates in a large water system with multiple inputs and outputs (Fluid Flow), determining pipe corrosion rates (Reaction Engineering), creating sampling plans (Unit Operations Lab), conducting mass balances and detecting potential point sources of contaminants (Mass and Energy Balances, along with knowledge of environmental chemistry concepts), identifying potential ways to mitigate further environmental and human impacts (Separations, Design), and projecting project costs (Engineering Economics). In addition to providing students the chance to review and build on previous course content, students expand their communication, problem solving and ethical decision making skills. As this case study requires chemical engineering specific knowledge, it is aimed at senior chemical engineering students. Our goal with this work is to cultivate a culture of technical and ethical responsibility among students, thereby preparing them to tackle real-world engineering challenges in their future careers.

521d An Engaging Course on Ethics, Professionalism, and Social Responsibility

Daniel Anastasio, Adam J. Nolte

Ethics and professional skills are critical to the safe and equitable practice of chemical engineering. Students, however, may view these topics as secondary to technical skills commonly taught in undergraduate curricula. In our department, we require a 1-credit professional practice course that has been historically used to expose students to contemporary issues and assess required ABET outcomes related to ethical reasoning. This course was typically maligned by students, who reported feeling disengaged and viewed attendance as an obligation rather than a learning opportunity.

To better promote student engagement with issues related to ethics and professionalism, we completely rebuilt the course so that now students regularly report strong interest in the content, citing that they view the course as highly applicable and even "fun". Utilizing case studies, game-based learning, and online environmental and socioeconomic data maps, we currently cover topics related to the intersection of chemical engineering with ethics, human psychology, sociology, and social responsibility. Course topics are framed around our obligations as chemical engineers to ethical decision-making, process safety, sustainability, and environmental justice.

In this presentation, we will share our course details and including lesson topics, assignments, and strategies for creating engaging assignments while minimizing the grading burden on faculty. This should be of particular interest for departments seeking effective methods for teaching and assessing topics related to ABET SO4, as well as broadening student perspectives on issues related to the environment, sustainability, and social equity.

521e Reflecting Changes in Communication Styles within Laboratory Courses

Beth Rundlett, Charles Stanier

Accreditation bodies like ABET, and universities have long recognized the need to equip students with effective communication skills. However, a significant shift towards conciseness and visual representation has transformed traditional communication methods. Engineering employers increasingly expect young professionals to be capable in various communication methods. These include customer presentations, policy drafting, public meetings, standard operating procedures, and graphical instructions, which extend beyond the conventional laboratory report that new recruits are expected to know. The changing styles, coupled with a perception that recent graduates lack the ability to articulate their thoughts spontaneously, call for a solution. These changes have significant implications for engineering education, particularly in laboratory courses where communication skills are emphasized.

Laboratory courses often serve as platforms for practicing communication. They emphasize the ability to communicate complex ideas, technical results, and intricate ideas, technical findings, and overarching concepts. However, evaluating this communication solely through lab reports, formal presentations, and lab notebooks may not sufficiently equip students for their professional lives. This work proposes potential methods to integrate diverse communication modalities into engineering education, particularly in the laboratory setting. These methods involve broadening assessments to include informal presentations, infographic posters, technical memos, standard operating procedures, graphical abstracts, and brief video presentations, in addition to traditional lab reports and formal presentations.

The shift away from written lab reports not only augments student skills but can also expedite assessment grading and bolster student accountability. For example, informal oral presentations cannot be plagiarized from the internet or peers. Instructors or teaching assistants can easily question all group members, thereby promoting individual accountability and reducing free-riding. Effective management of free riding by the instructional team can boost morale and student satisfaction. A second example is that grading a single-page infographic poster is considerably quicker than grading a 20-page lab report, allowing for the assessment to be individualized rather than a team-based approach. Lastly, it is vital for students to verbally express their experiences in addition to writing about them, familiarizing them with the terminology associated with chemical engineering and preparing them for interviews and roles in research and production environments.

Think like an expert". An engaging in-class activity for sustainable engineering courses

Joaquin Rodriguez, Department of Chemical and Petroleum Engineering,
University of Pittsburgh

Team building games are effective strategies to develop communication, teamwork, and leadership skills [1]. They rank among the top skills demanded by industry for new professionals. In-class activities developing these skills in curricular college courses enrich the formation of students in this direction and offer great potential to engage them in effective collaboration and reflect on their experiential-based learning. The activity reported here aimed to provide students with an experience to realize the value of communication, negotiation, leadership, and teamwork, based on previous technical knowledge acquired during the course, but open to uncertainties and opinion. The activity was developed and tested in an elective course for junior and senior students on green chemical engineering and sustainability [2]. The activity provided students with the attractive feature of approaching experts' qualified opinion about main topics explored in the course. It was based on a yearly survey that the life insurance institution AXA has been performing since 2014. AXA requests from experts and general population around the world to rank the top 10 future world risks (including themes like climate change, cybersecurity, energy risks, pollution, etc.). For the 2023 edition (published in October), they obtained responses from 3,500 experts and over 22,000 individuals (public). The combination of the final two lists resulted in 12 top future risks (8 were common in the expert and public lists). Students were presented with this final selection of 12 themes and were asked to build their own ranking, without having information on the reported rankings. Students were given 10 minutes to select their top 10 choices individually, with no communication to other students. After that first selection, they were paired and requested to decide on a common ranking by consensus or heads-and-tails as alternative method. Students were given 10 minutes for this second phase. In a third round the pairs were arranged in teams of 5-6 members (the same teams they had been working projects for the course during the entire semester), and requested to get the team's top-10 list by consensus, vote or heads-and-tails, always favoring consensus, and recording the decision method, for another 10 minutes. Finally, the entire class was assigned with the task of selecting the class' top-10 list for another 10 minutes. Students were invited to take leadership roles in order to guarantee the completion of the list on time. The final part of the activity asked students to compare each list (individual, pair, team, class) with the experts' selection. Points were awarded or subtracted depending on the proximity of the student selections to the experts' rank. Results showed a consistent improvement with increasing dialogue and negotiation, from the individual step to the whole class performance, with a net improvement of 4 points (in the scale of 1-12) by this evolution in teamwork. All the student lists averaged better than the public list when compared with expert selection, as a more informed segment of the population, derived from their formation in the course. The activity was executed as the concluding class for the course, provided a wrap-up opportunity for selected main topics, and clearly demonstrated the efficacy of collaboration, communication, and teamwork.

1. Introduction

This paper elaborates on the presentation given at the Teaching Professional Skills session of the Education Division at the AIChE 2024 Annual Meeting (San Diego, CA 10/30) (Slide 1). It describes the results with an engaging activity developed for a course on "Green Chemical Engineering and Sustainability" attended by junior and senior chemical engineering students in the fall 2023. It was designed for a 50-minutes session in person. However, the structure can be easily adapted to other contexts and programs provided that a ranked list of issues, topics, or items prepared by experts is available and of the interest of students to satisfy desired outcomes. The activity is intended to develop or reinforce communication and teamwork skills, handling opinions in a consensual approach, managing decisions on time, and exposing students to some special area of knowledge (i.e., the main future risks for society in this case).

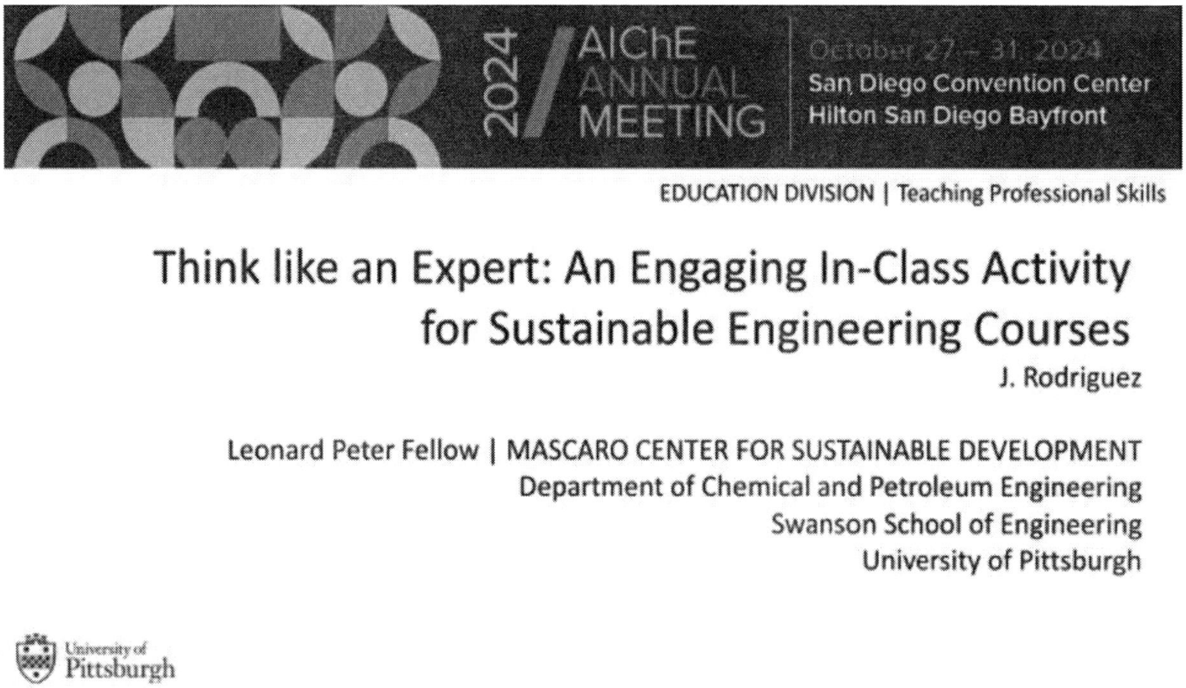

Slide 1. A presentation on an in-class activity to engage students in a topic previously analyzed by experts and comparing how students approach experts' selection of ranked topics by advancing dialogue and teamwork.

2. Teambuilding games for engaging in-class activities

This activity could be classified among teambuilding games (TBG). Teambuilding games are at the intersection of the "Science of Team Science (SciTS)" and "Game Based Learning (GBL)". These are fields of increasing development gaining popularity in colleges, businesses and industry. They address the training and development of skills for effective teamwork in safe environments for testing and reflecting on attitudes, biases, character traits, strategies, communication patterns, and other features influencing team performance (Slide 2)[1]

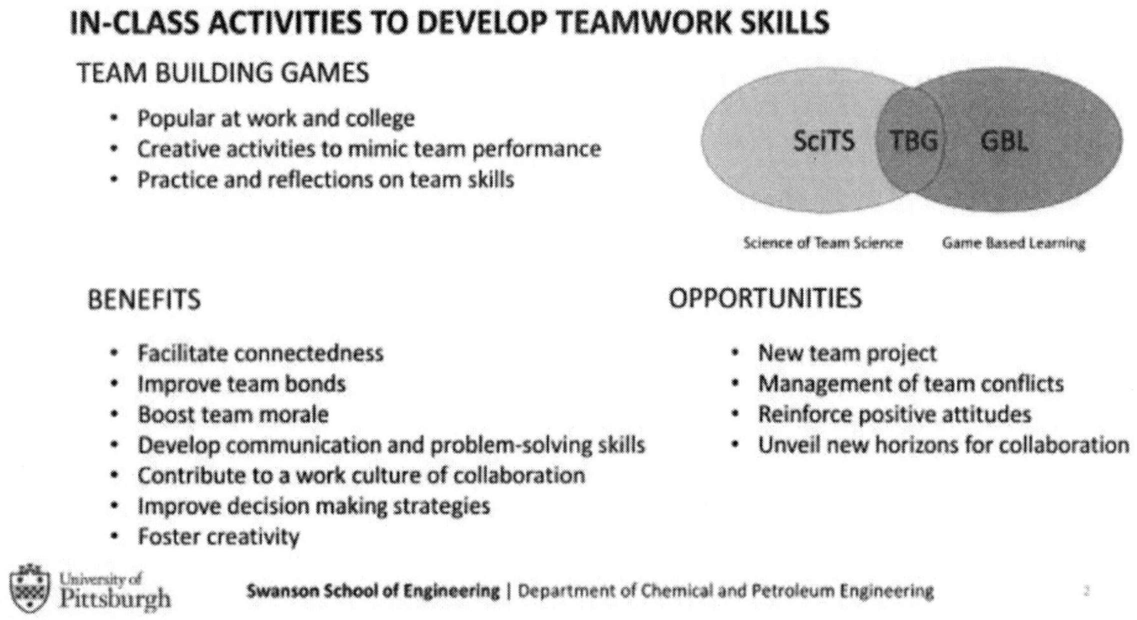

Slide 2. Teambuilding games lay at the intersection of two areas of increasing research and popularity in colleges, industry and businesses, “Science of Team Science" and “Game Based Learning"

3. A team building-based in-class activity in a sustainability focused chemical engineering course

This specific activity was designed for the last lecture on an elective course on “Green Chemical Engineering and Sustainability", attended by 46 junior and senior chemical engineering students in the fall 2023 (Slide 3)[2]. The course was meeting three times a week in 50-min session each. The instructor, with occasional participation of invited speakers, conducted the sessions. The classroom environment was open for discussion and in-class analysis of a broad range of topics on sustainability from a chemical engineering perspective, including many in class activities for engaging students with special topics and fostering an educational space for collaboration (Slide 4).

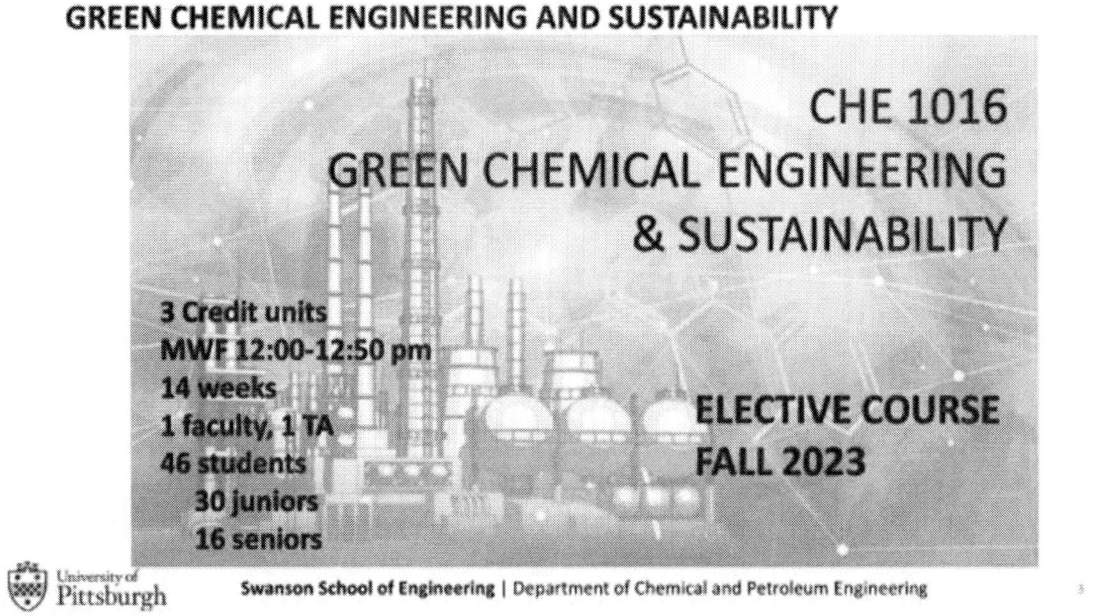

Slide 3. An elective course on "Green Chemical Engineering and Sustainability" provided the opportunity for deploying the activity "Thinking like an expert"

Slide 4. The classroom environment

The course “Green Chemical Engineering and Sustainability" was structured mainly on the textbook “Green Chemistry and Engineering", by Dr. Conchita Jimenez-Gonzalez and Dr. David J. Constable, enriched with many sources on current issues in sustainability (Slide 5). One distinctive feature of the course was the exposure of students to 14 experts from different areas (government, industry, academia, business, NGOs), both in-person and virtual. Students had the opportunity to interact with these experts in the classroom, and they were requested to provide graded short reflections on the learnings from each participant speaker.

In addition, student teams (4-6 members) developed an open-ended self-selected project on sustainability from a chemical engineering perspective, with options for technology improvement, community enrichment, research initiatives or policy proposals. Project deliverables included a brief 5-min presentation and a poster. These deliverables were displayed in a public event (“CHE Sustainability Day 2023”) with judges from industry and academia. This was another occasion where students had the opportunity with interact with experts in various fields.

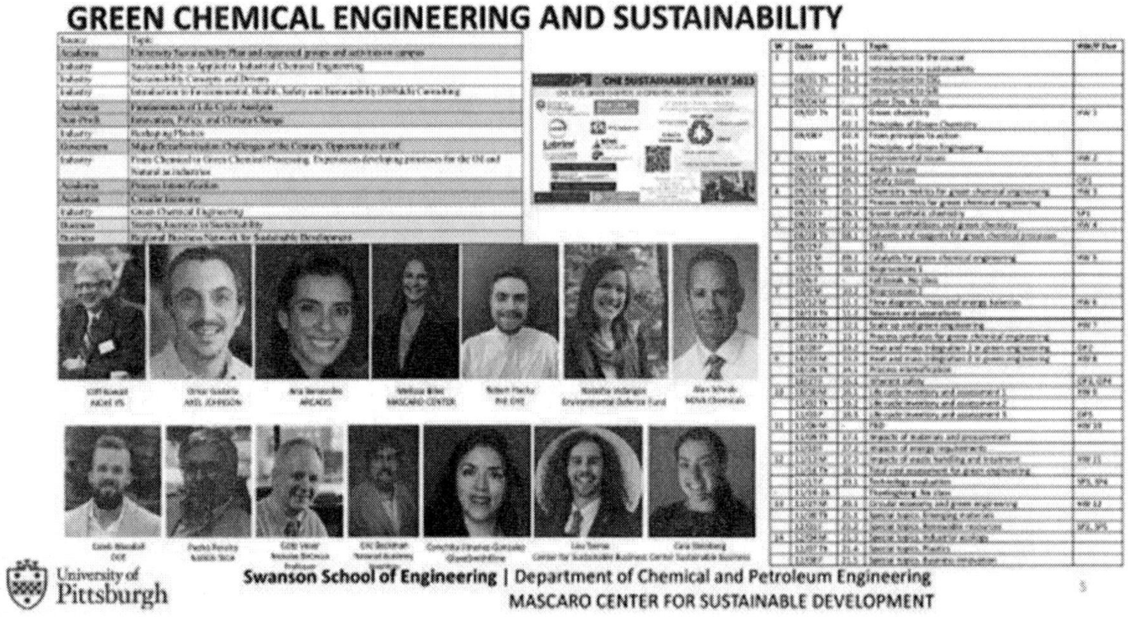

Slide 5. The course “Green Chemical Engineering and Sustainability" provided several opportunities for students to interact with experts from various areas (research, technology, policy-making, industry, academia, NGO).

4. A documented expert assessment on main future risks for humankind

This activity took opportunity of the annual publication of “Future Risks" in the 10th edition (Report 2023) by AXA[3] (Slide 6). This multinational financial company assesses the risk environment for the future by surveying annually about 3,500 experts from 50 countries, and over 20,000 individuals in 15 countries about the top ranked risks for the future. Participants are presented with a former list of about 25 risks for them to rank the top 10. For the 2023 report, the experts' list and the general population's list agreed on 8 risks, though at different ranks, and 4 additional risks made to the list of experts or public. The 1combination of the two lists results in 12 top ranked risks as presented in slide 6.

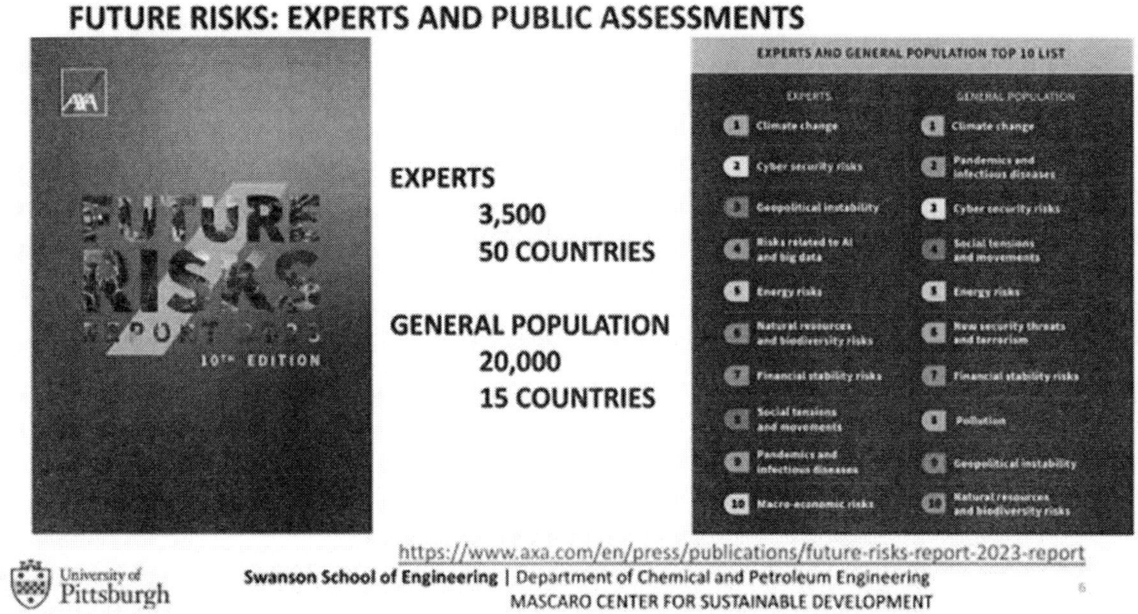

Slide 6. Top 10 future risks by experts and general public compiled by AXA in 2023.

5. The structure and development of the in-class activity

The instructor of the course presented students with this list of 12 future risks, but the students were not aware of the selection by experts or public. Some of these risks were considered extensively during the course (i.e., climate change, energy, natural resources and biodiversity depletion, pollution) while others were not included.

In the first iteration of the activity, students were required to work out the list of 12 future risks and select their 10 top-ranked risks top-down. They were given 10 minutes for the selection using the format displayed in the slide 7.

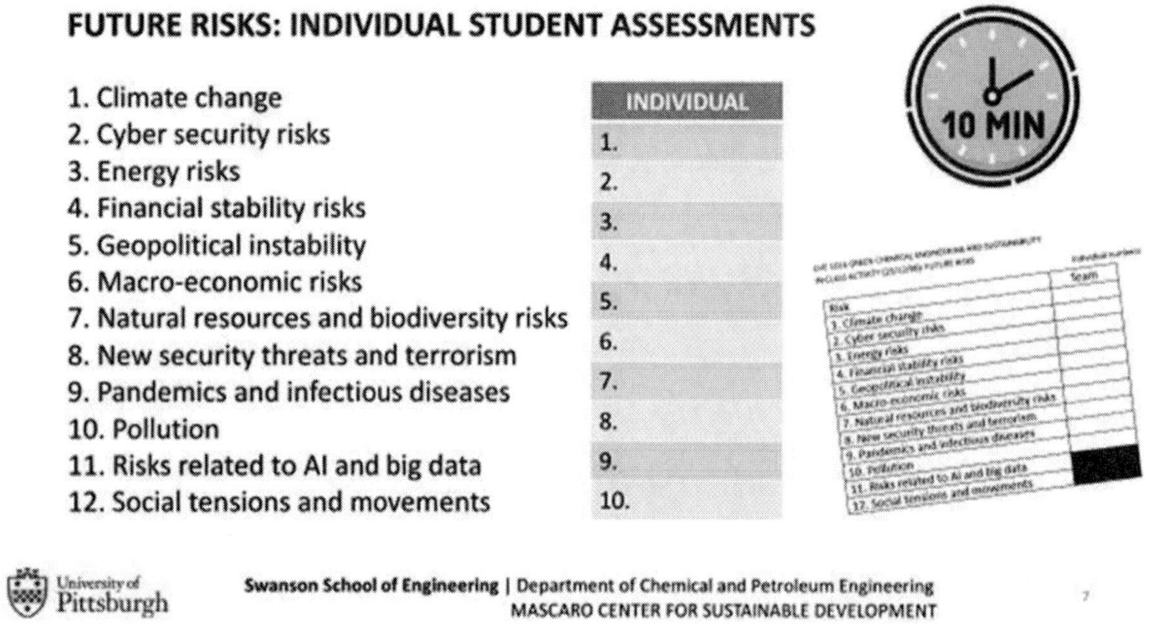

Slide 7. In the first iteration of the activity, students were required to top-down rank the 10 most critical risks for the future of mankind based on a previous list complied by AXA.

For that final day of classes, 31 students were attending the activity. The results for their selection of 10 top ranked risks is displayed in slide 8. About one-third of the students selected "climate change" as the top risk, followed by "natural resources and biodiversity risks" (%), and "geopolitical instability" (%), but other 6 risks made it to the top of the selection for some students. Singularizing the "climate change" risk, despite being the dominant selection for one-third of students, it ranked down, even to the seventh position for some students. A first insight on these results is the broad range and diverse selection from students who have been sharing the same college experience for three or four years.

FUTURE RISKS: INDIVIDUAL STUDENT ASSESSMENTS

Student	1	2	3	4	5	6	7	8	9	10	11	12	13	14	15	16	17	18	19	20	21	22	23	24	25	26	27	28	29	30	31
1. Climate change	1	5	1	1	1	8	5	4	5	7	5	7	1	7	7	10	11	7	1	1	1	5	2	9	1	8	1	2	7	1	1
2. Cyber security risks	7	4	7	7	7	9	2	7	2	5	3	5	7	10	10	1	2	12	3	6	3	3	1	6	5	2	7	9	1	10	2
3. Energy risks	10	3	10	10	3	1	3	2	1	10	7	12	3	1	1	7	1	5	10	2	7	2	8	11	2	7	3	1	9	9	3
4. Financial stability risks	5	12	5	4	12	4	1	5	10	1	6	3	4	3	5	5	5	11	4	10	4	1	9	1	3	1	8	7	10	5	10
5. Geopolitical instability	9	1	3	3	11	7	8	8	11	12	10	8	6	5	3	8	8	1	7	8	12	7	10	7	7	4	9	10	12	3	8
6. Macro-economic risks	3	7	11	5	10	5	10	9	7	8	1	9	5	4	6	11	3	3	6	5	6	6	4	4	6	12	10	3	5	12	4
7. Natural resources and biodiversity risks	8	6	6	12	8	11	12	1	6	4	2	1	2	6	9	3	12	4	5	7	10	10	3	12	4	10	4	8	3	7	7
8. New security threats and terrorism	11	10	8	8	2	10	7	3	4	9	4	10	8	12	12	9	7	6	12	9	5	11	6	10	11	5	5	4	8	8	5
9. Pandemics and infectious diseases	6	9	9	2	5	6	6	10	3	3	8	11	10	8	4	4	4	10	8	3	2	4	5	5	10	6	8	11	6	2	12
10. Pollution	2	8	12	9	6	12	11	11	9	2	9	4	9	9	11	12	9	9	9	4	9	9	7	8	8	3	2	12	4	11	11
11. Risks related to AI and big data																															
12. Social tensions and movements																															

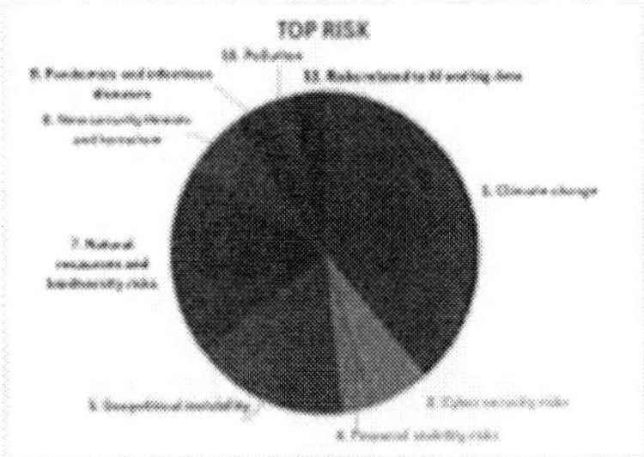

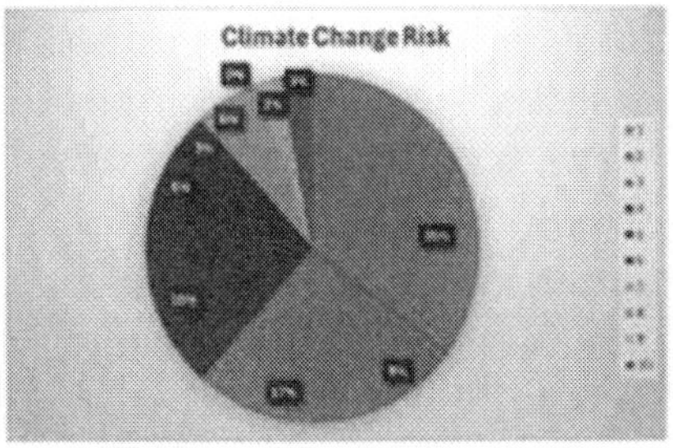

University of Pittsburgh

Swanson School of Engineering | Department of Chemical and Petroleum Engineering
MASCARO CENTER FOR SUSTAINABLE DEVELOPMENT

Slide 8. Results from the first iteration of the activities with individual students selecting their own top-10 list of future risks.

For the second iteration of the activity students were paired with a member of the team working the sustainability project for the class. They were given another 10 minutes to revisit their lists and come out with a consensus list by discussing their choices. In case of not reaching agreement at any position they were offered to take the decision flipping a coin and recording the number of times that they reached the decision this way (Slide 9). Results are presented in slide 10. In this iteration, the "top risk" selection got reduced to three options (the dominant options from the previous iteration) showing a convergent selection reached mainly by consensus. It can be noted that 5 pairs had the recourse of flipping coins for reaching decision in the frame of the 10-min period, though it is not specified for what positions in the rank. In the same way, the consideration of "climate change" as a top risk was distributed in four positions instead of 8 for the first iteration. These results suggest how a dialogue in a safe environment for discussion based on a common background for education can produce effective consensus, even under constraints like time limits.

FUTURE RISKS: STUDENT PAIR ASSESSMENTS

1. Climate change
2. Cyber security risks
3. Energy risks
4. Financial stability risks
5. Geopolitical instability
6. Macro-economic risks
7. Natural resources and biodiversity risks
8. New security threats and terrorism
9. Pandemics and infectious diseases
10. Pollution
11. Risks related to AI and big data
12. Social tensions and movements

PAIR
1.
2.
3.
4.
5.
6.
7.
8.
9.
10.

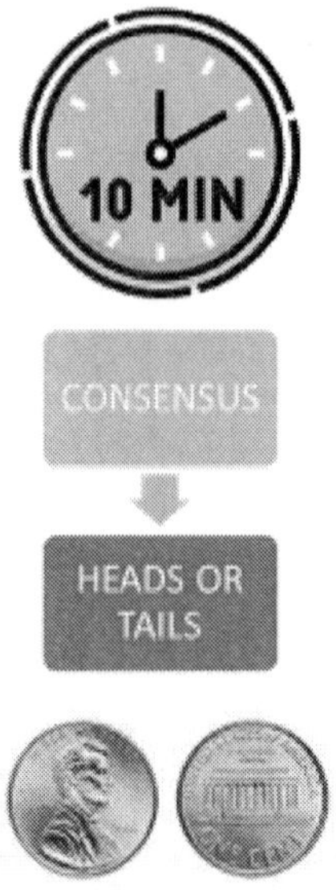

University of Pittsburgh   **Swanson School of Engineering** | Department of Chemical and Petroleum Engineering
MASCARO CENTER FOR SUSTAINABLE DEVELOPMENT   9

Slide 9. In the second iteration of the activity, students were paired with one teammate to produce a consensus list, including the option of flipping a coin if dialogue did not lead to agreement.

FUTURE RISKS: STUDENT PAIR ASSESSMENTS

Pair	1	2	3	4	5	6	7	8	9	10	11	12	13	14	15
Students	1,6	2,3	4,5	7,10	8,9	11,12	13,14	15,16	17,18	19,20	21,22	23,24	25,26	27,28,29	30,31
1. Climate change	1	1	1	5	5	5	1	7	3	1	1	5	1	7	1
2. Cyber security risks	9	5	7	7	2	3	7	1	11	3	5	1	2	1	2
3. Energy risks	7	3	8	4	7	7	3	10	7	6	3	6	5	9	3
4. Financial stability risks	8	7	10	1	1	12	10	5	1	10	7	9	8	3	10
5. Geopolitical instability	10	4	12	6	4	6	5	6	12	4	2	2	4	10	5
6. Macro-economic risks	5	6	11	12	10	10	6	3	3	7	6	4	10	2	4
7. Natural resources and biodiversity risks	11	10	8	10	9	1	4	9	2	5	4	8	3	8	7
8. New security threats and terrorism	3	11	4	3	11	8	8	11	8	12	10	7	7	12	8
9. Pandemics and infectious diseases	6	12	5	2	6	11	12	4	4	8	11	10	11	4	12
10. Pollution	4	9	2	9	8	9	9	12	9	9	12	11	6	6	11
11. Risks related to AI and big data															
12. Social tensions and movements															
Coin flips	2	0	0	0	1	2	0	2	0	0	0	3	0	0	0

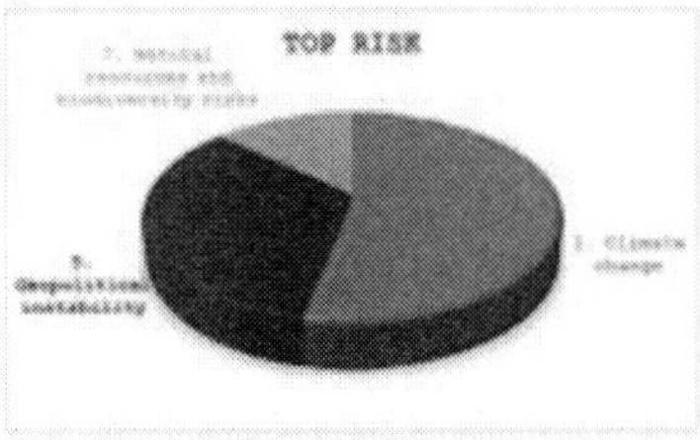

University of Pittsburgh   **Swanson School of Engineering** | Department of Chemical and Petroleum Engineering
MASCARO CENTER FOR SUSTAINABLE DEVELOPMENT   10

Slide10. Results from the second iteration of the activity where students were paired with one teammate to produce a consensus list.

In the third iteration of the activity, students were gathering with teammates for the course project (4-6 members) and tasked with the same assignment of producing a consensus list in 10 minutes. This time they were given the additional option of considering voting before flipping coin to agree on a debated choice, and asked to record the number of times they made use of these two options (Slide 11).

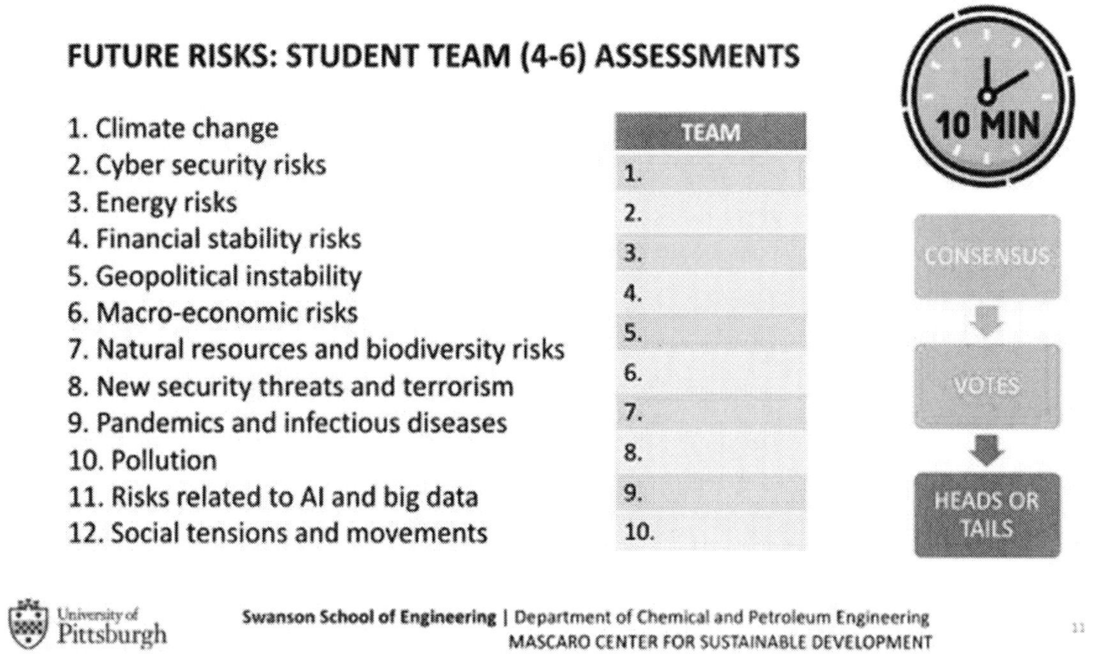

Slide 11. In the third iteration of the activity, students worked in the team for the course project to decide on a consensus list, recurring to voting or flipping coin if needed.

Results are presented in slide 12. The teams reached consensus in all choices but one for a single team, showing a high level of convincing discussions and agreement, following the previous informed consideration of the choices in the individual and paired stages of the activity. One group brought to the top of the list one risk (2. Cyber security risks) not previously listed in that position by the pairs, but by individual members in that team.

FUTURE RISKS: STUDENT TEAM (4-6) ASSESSMENTS

Team	1	2	3	4	5	6
Students	1,2,3,4,5,6	7,8,9,10	11,12,13,14,15,16	17,18,30,31	19,20,21,22,23,24	25,26,27,28,29
1. Climate change	1	5	7	1	1	2
2. Cyber security risks	7	7	5	5	5	1
3. Energy risks	3	2	1	3	3	7
4. Financial stability risks	10	1	3	7	6	3
5. Geopolitical instability	5	4	10	2	7	10
6. Macro-economic risks	11	10	6	12	4	8
7. Natural resources and biodiversity risks	8	8	12	11	2	9
8. New security threats and terrorism	9	9	9	8	10	5
9. Pandemics and infectious diseases	4	12	11	9	12	12
10. Pollution	6	3	8	4	8	6
11. Risks related to AI and big data						
12. Social tensions and movements						
Votes	0	1	0	0	0	0
Coin flips	0	1	0	0	0	0

University of Pittsburgh

Swanson School of Engineering | Department of Chemical and Petroleum Engineering
MASCARO CENTER FOR SUSTAINABLE DEVELOPMENT

12

Slide 12. Results from the third iteration of the activity with students working on teams (4-6 members) to produce consensus lists.

Finally, in a fourth iteration of the activity, the whole class was tasked with the assignment of producing a consensus list in the same period of 10 minutes. It was interesting to observe the dynamics emerging in this stage of the activity. Some students were taking the lead at proposing a choice for a position, starting with the top one. It came naturally to develop the list top down, with no further instruction for that method. Sometimes some students proposed simultaneously different choices, leading to at least three occasions to take a vote. At times, there was silence and inactivity to move forward and the instructor was asking for initiative and leadership. No decision was taken by flipping coins. However, no records were taken on the development of the activity. Results are presented in slide 13.

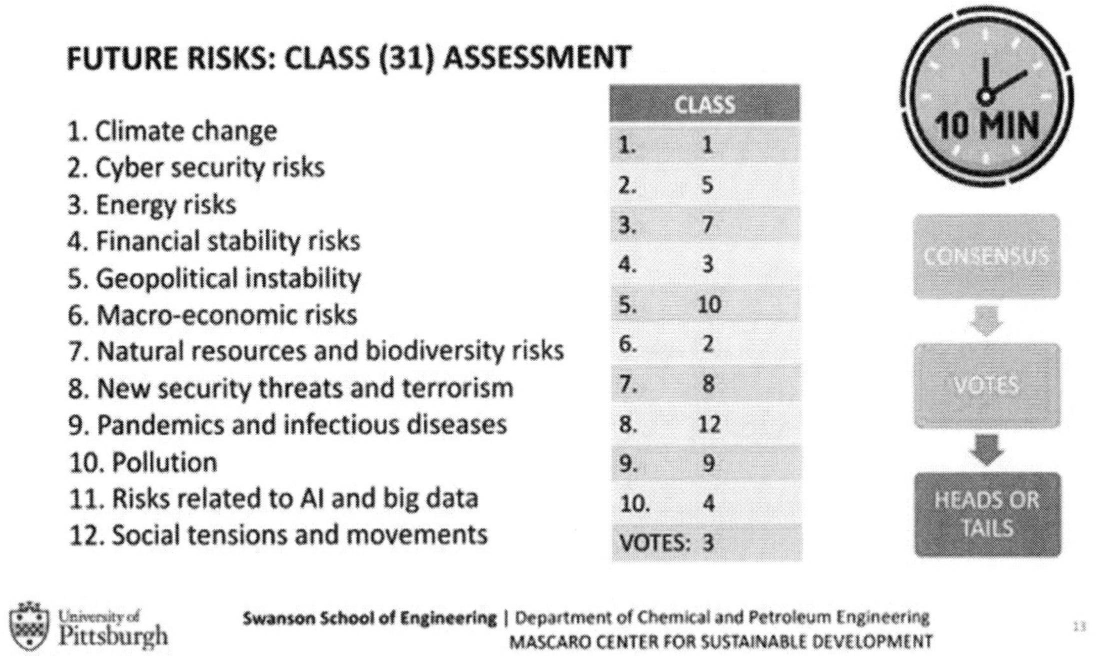

Slide 13. Results from the fourth iteration of the activity with students working together as a class to produce a consensus list.

6. Comparison with expert assessment

The final part of the activity consisted in showing students the survey results from experts and general public. In addition, to approach a quantitative assessment of the eficacy of the activity, an arbitrary scoring scale was proposed to illustrate the differences with respect to the assessment by experts, as reported in slide 14. Scores are based on the proximity of a given choice to the corresponding position selected by the experts, i.e., if matching the same position, a score of +1 is given, if it is one position apart, a score of +0.5 is give, at the low end, if the choice is 4 or more positions apar, a -1 score is computed. Individual scores can be added for an overall estimation of how close are the selections to the expert opinion. A preliminary exercise consisted in showing the comparison of the AXA survey with public opinion as reported in slide 15. The first five positions show a difference of +0.5 out of a potential +5 (to reflect perfect matching with expert assessment), extending the analysis to the 10 positions the overall estimation decay to -3.5.

THINKING LIKE EXPERTS: SCORING PROCEDURE

1. Climate change
2. Cyber security risks
3. Energy risks
4. Financial stability risks
5. Geopolitical instability
6. Macro-economic risks
7. Natural resources and biodiversity risks
8. New security threats and terrorism
9. Pandemics and infectious diseases
10. Pollution
11. Risks related to AI and big data
12. Social tensions and movements

EXPERTS	
1.	1
2.	2
3.	5
4.	11
5.	3
6.	7
7.	4
8.	12
9.	9
10.	6

SCORING
- If you hit the same position = +1
- If your choice is ± 1 position = +0.5
- If your choice is ± 2 position = +0
- If your choice is ± 3 position = -0.5
- If your choice is ± 4+ position = -1

University of Pittsburgh   Swanson School of Engineering | Department of Chemical and Petroleum Engineering
MASCARO CENTER FOR SUSTAINABLE DEVELOPMENT   14

Slide 14. The experts' assessment and the scoring scale.

THINKING LIKE EXPERTS: PUBLIC OPINION

1. Climate change
2. Cyber security risks
3. Energy risks
4. Financial stability risks
5. Geopolitical instability
6. Macro-economic risks
7. Natural resources and biodiversity risks
8. New security threats and terrorism
9. Pandemics and infectious diseases
10. Pollution
11. Risks related to AI and big data
12. Social tensions and movements

EXPERTS		PUBLIC		SCORE	
1.	1	1.	1	1.	+1
2.	2	2.	9	2.	-1
3.	5	3.	2	3.	+0.5
4.	11	4.	12	4.	-1
5.	3	5.	3	5.	+1
6.	7	6.	8	6.	-1
7.	4	7.	4	7.	+1
8.	12	8.	10	8.	-1
9.	9	9.	5	9.	-1
10.	6	10.	7	10.	-1
				Total:	-3.5

University of Pittsburgh   Swanson School of Engineering | Department of Chemical and Petroleum Engineering
MASCARO CENTER FOR SUSTAINABLE DEVELOPMENT   15

Slide 15. Comparison of the AXA survey of public opinion and the assessment by experts

Finally, the scores from students at the different stages of the activity were computed, with results illustrated in Table 1 and slide 16. Though results should be taken as merely indicative of a potential pattern, they show that students at any stage reach results closer to the expert assessment than the public opinion, maybe confirming the advantage of a formal education in several of the topics addressed in the survey. More interestingly, they show and increasing approach to the experts' selection from the individual stage, to the pairing selection, the team selection, and finally the class assessment, illustrating the advantages of dialogue, inclusiveness, leadership, and knowledge sharing. They also show an increasing reduction on the dispersion of results as the activity evolves from individual to pairs, and to teams. It cannot be ignored also the gap still remaining with expert evaluation, providing some indication of the value of expertise.

Table 1. Summary of scores by the various groups

Group	Score
Experts (reference)	+10
Class	+ 1.0
Team	- 1.7
Pairs	- 2.4
Individuals	- 2.6
Public	- 3.5

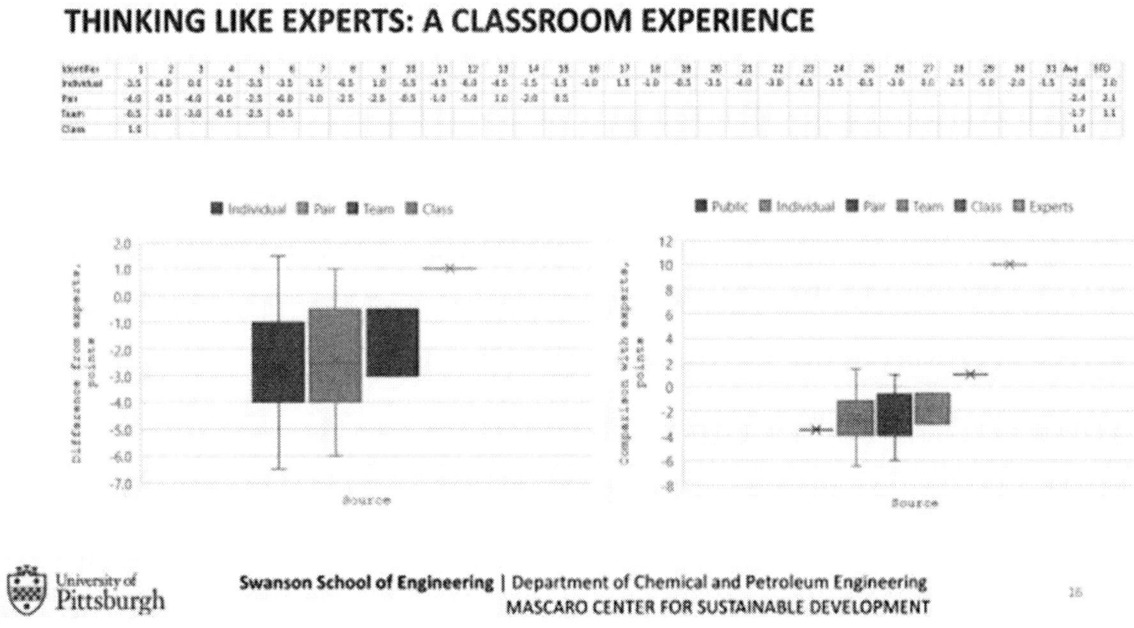

Slide 16. Comparison of scores from students and experts (left) and including public opinion survey (right).

7. Takeaways and future work

As presented in slide 17, a major takeaway from this in-class activity was the exposure of students to a critical topic on main future risks in a conversational approach that allows them to explore their perceptions and share them with classmates while attempting to rank them. The activity engaged students in communicating their opinions and openly provide and receive feedback in a safe format simulating a decision-game with realistic constraints. Students were able to explore teamwork skills at reaching consensus decisions. The activity provided a quantitative assessment that illustrated the advantages of pursuing consensus decisions based on informed opinion.

This activity could certainly have been improved if support were available to record and further analyze the dynamics of pairs, teams, and class to reach the goals. The corresponding narratives could provide opportunities to examine team behaviors and relate them to performance. The activity could have also benefited from the support of a digital platform to display results on real time, speeding calculations and graphics, adding a more immediate analysis of team performance. These and other improvements are to be considered in further editions of the activity and recommended for application in similar exercises, as emphasized in slide 18.

TAKEAWAYS

This "Think like experts" activity provided an engaging experience for a 50-min session stimulating students to think about the future, from the perspective of main risks.

The activity fostered the development of communication skills, handling opinions in a team, leadership and team-building.

The activity illustrated the advantages of dialogue and consensus to reach more balanced and reliable assessments on topics of concern.

University of Pittsburgh — Swanson School of Engineering | Department of Chemical and Petroleum Engineering
MASCARO CENTER FOR SUSTAINABLE DEVELOPMENT 17

Slide 17. A selection of major takeaways from the activity

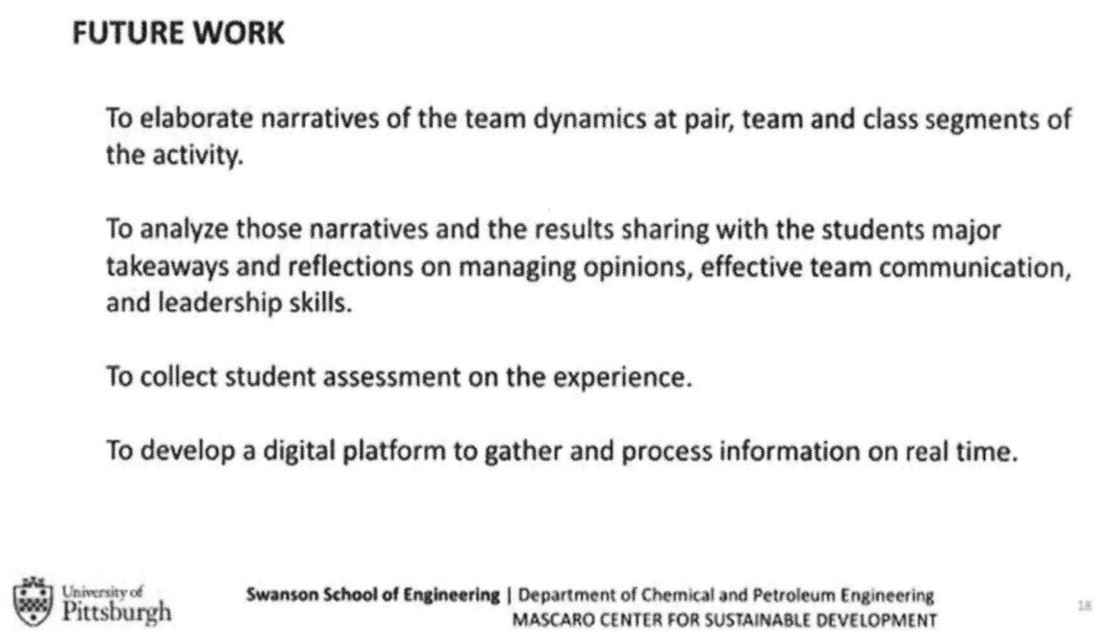

Slide 18. Suggestions for further work in new editions of this activity or similar exercises.

Acknowledgements

The author is grateful for the Leonard Peters Faculty Fellowship by the Mascaro Center for Sustainable Innovation at the University of Pittsburgh for the support, encouragement and vision to foster sustainability in higher education.

References

[1] Rodriguez, J., Baled, H., & McMahon, M. (2023). Team Building Games to Reinforce the Training of Chemical Engineering Students in Team Skills Based on Collaboration Leadership. In *2023 ASEE Annual Conference & Exposition Proceedings*. American Society for Engineering Education. doi: 10.18260/1-2--44021.

[2] Rodriguez, J. (2024). An elective course in Green Chemical Engineering and Sustainability. In *2024 ASEE North Central Section Conference Proceedings*. American Society for Engineering Education. doi: 10.18260/1-2--45596.

[3] AXA Future Risks Report 2023. https://www.axa.com/en/news/2023-future-risks-report

521g Integrating Professional Skills into Graduate STEM Education

Victor M. Ugaz

Equipping students with professional skills is critical in preparing engineering undergraduates for successful careers. There is also growing recognition of the vital role these skills play in ensuring the success of graduate students, owing to the increasingly competitive and interdisciplinary nature of the modern workforce. But despite this acknowledged importance, the integration of professional skill development into graduate curricula, especially in a manner that addresses the distinct needs of graduate students, remains underexplored. Here, we show how professional skills have been introduced to prepare students in the Master of Biotechnology (MBIOT) program at Texas A&M University, a professional STEM Master's (PSM) program enabling students to receive critical interdisciplinary training that includes strong business, professional development, and experiential components. Here, we describe three strategies to embed professional skills within the graduate curriculum. First, we embedded new activities and deliverables into a core seminar course framed in terms of a personal business model described in the book *Business Model You: A One-Page Method for Reinventing Your Career by Osterwalder et al.* These activities encourage students to undertake a professional inventory and articulate their career interests and objectives through personal statements. This not only aids students in defining their professional trajectory but also acquaints them with foundational entrepreneurship principles. Second, we embedded a new bioethics module inspired by the 2018 best-selling book *Bad Blood: Secrets and Lies in a Silicon Valley Startup* by John Carreyrou, chronicling the rise and fall of the biotech startup Theranos and its founder, Elizabeth Holmes. This module, delivered in a 'book club' format, promotes engagement and discussions about vital ethical considerations in biotech ventures. Third, recognizing the importance of effective communication in science and engineering fields, the curriculum now includes a dedicated module aimed at refining students' abilities to convey scientific concepts to diverse audiences through compelling storytelling. These curriculum innovations have bolstered the students' professional skills, contributing to the broader objective of preparing well-rounded engineering professionals.

521h Impacts of the Hybrid Format on the Attendee Demographics and Carbon Footprint of Scientific Conferences

Ronald J. Vogler, Bryson S. Dawson, Mason F. Gilb, Hyeonji Oh, Manish Kumar

Technical conferences are critical to the professional development of STEM students and professionals at any career stage, allowing participants to engage in professional development, continuous learning, and networking. During the COVID-19 pandemic, the prevalence of the virtual conference format greatly increased and was shown to have advantages such as lower attendee cost, lower conference carbon footprint, and the increased representation of individuals from underrepresented genders and institutions. Despite these benefits, many attendees prefer a conference with an in-person component as certain conference activities, particularly networking, were limited by the virtual format. One approach to provide attendees with in-person conference activities while reaping some benefits of the of virtual conference format could be to conduct "hybrid" conferences that involve the simultaneous participation of in-person and virtual attendees. In this work, the impact of the hybrid format was investigated for four conferences: the 2022 North American Membrane Society Conference (NAMS 2022), the 16th Deep-Sea Biology Symposium (16DSBS), and the 240th and 241st American Astronomical Society Conferences (AAS 240 and AAS 241), each with different percentages of participants that attended online. For the NAMS 2022 conference, the hybrid format resulted in a 31.5% reduction in carbon footprint (~154 Tonnes CO2e) compared to a hypothetical case where all attendees went to the conference in person. Online attendance also provided a more financially accessible mode of conference attendance; the average online attendee paid only $142 to attend NAMS 2022, whereas the cost for the average in-person attendee was $1877. Furthermore, the online attendee population at NAMS 2022 contained a larger percentage of international attendees, women, and undergraduates than the in-person attendee population, suggesting that the offering of an online attendance option bolstered the involvement of these groups. Similar analyses will be discussed for the 16DSBS, AAS 240, and AAS 241 conferences to further determine the impact of the hybrid format on attendee demographics and conference carbon footprint as well as the magnitude of these impacts in relation to the percentage of participants that attended online.

AUTHOR INDEX

AIChE
120 Wall Street, FL 23
New York, NY 10005-4020

ISBN 979-8-3313-1661-7

CURRAN ASSOCIATES INC.
proceedings.com